Contractor's
Business
Handbook

CONTRACTOR'S BUSINESS HANDBOOK

Michael S. Milliner

Contributing Authors:
Donald Butt, Jr.
Keith Fetridge

ACCOUNTING

FINANCE

TAX MANAGEMENT

COST CONTROL

R.S. Means Company, Inc.

R.S. MEANS COMPANY, INC.
CONSTRUCTION CONSULTANTS & PUBLISHERS
100 Construction Plaza
P.O. Box 800
Kingston, Ma 02364-0800
(617) 585-7880

© 1988

This book was edited by Mary Greene and Neil Smit. Typesetting was supervised by Helen Marcella. The book and jacket were designed by Norman Forgit. Illustrations by Carl Linde.

Printed in the United States of America

10   9   8   7   6   5   4   3   2   1

Library of Congress Cataloging in Publication Data

ISBN 0-87629-105-1

To Debbie, Lucas, and Timothy who suffered with me as I worked on this book weeknights, weekends, holidays, and vacations—and still missed every deadline.

CONTENTS

FOREWORD

This book is designed for contractors interested in establishing a firm financial structure for their businesses, and planning a sound program for future growth and development. The book presents the contractor's most important tool for financial management, an efficient Financial Control System. Chapters 2 through 8 cover the seven key elements of the system, and outline specific methods for their implementation. A hypothetical construction company is used throughout the book to illustrate an effective financial control system.

After an introductory chapter on the trends and special demands of today's construction industry, Chapter 2 begins to outline the financial control system by presenting the accounting procedures and strategies that are most important to a construction firm.

A properly designed and operated accounting system serves as the basis for the firm's ability to analyze, forecast, and minimize costs. Chapters 3 and 4 are devoted to financial analysis and forecasting. They define the contractor's financial structure and explain how to use trend analysis and several fundamental ratios to review a company's progress toward its goals.

Prudent cash management may be one of the most important goals of the contractor, who must have adequate working capital in order to stay in business. Chapter 5 describes the cash flow cycle, budgeting, and techniques for raising cash as well as accelerating cash receipts and slowing down disbursements.

Chapter 6, "Asset Management," is devoted to the best use of current and capital assets, and includes a section on the contractual means available to protect the firm's assets.

Chapter 7, "Debt Management," explains how to calculate the cost of financing and reviews the considerations of short and long term debt. There are also guidelines for dealing with lending institutions and a comparison of leasing versus purchasing.

Foresight and strategy must also be applied to the management of tax obligations if a firm is to have the best chance of prospering. The tax implications of the corporation and other organizational structures are compared in Chapter 8. This chapter also covers methods of income recognition and accounting, treatment of fixed assets, and many other tax considerations.

Acknowledgements

This book would never have been completed without the patience and contributions of many individuals. I wish to express my sincere gratitude to each and every one of them.

I would like to thank Don Butt for his timely and superb efforts on Chapters 6 and 7 and for his contributions to Chapter 5. Keith Fetridge did a wonderful job covering tax management in Chapter 8. I would also like to thank Dick Werner, of Shaker Computer Products, for reviewing and editing Chapter 9, "Automated Financial Applications", as well as my associate and friend, John Dallavalle, who advised me on many technical points.

My office staff was wonderfully supportive throughout this endeavor. Mike Martin and Linda Stefanon, past and current controllers for Milliner Construction, provided technical guidance and actual text in some cases. Lisa Brandenburg and Lisa Hildebrand both assisted in typing and preparing the figures. In addition, Nancy Rice, my dedicated and loyal administrative assistant, helped me tremendously throughout the entire effort. Dean Ventola assumed many of my business responsibilities, allowing me the time to work on this book. My hat is off to them all.

Finally, a special salute to Mary Greene, my manuscript editor at R.S. Means Company, who withstood my craziness more than anyone else.

Chapter One

FINANCIAL CONTROL IN CONSTRUCTION

Chapter One

FINANCIAL CONTROL IN CONSTRUCTION

Today's Construction Industry

The fortunes of virtually all U.S. business organizations are tied to today's complex and unpredictable business environment. Excessive regulations, economic instability and increasing liability represent just a few of the "loose cannons" that must be anticipated and dealt with almost daily. In addition to these external factors, managers must cope with a multitude of internal factors, such as cost control, productivity, and quality control.

Perhaps no industry epitomizes the convergence of these multiple factors more than real estate development. The very nature of the end product—functionally diverse buildings—requires developers and contractors to maneuver through an often undefined maze of regulations, market risks, and external participants in order to bring about the desired results.

Figure 1.1 shows, in simplified form, the steps typically involved in the development process. Each of these steps requires extensive expertise, effort, and time. Furthermore, contractors must establish and concurrently operate temporary "factories" at any number of sites, where a variety of resources are brought together in the actual construction of buildings. Considering the complexity of this business, it is no wonder that more than 50% of the new construction firms in this country fail within five years, with most of those going under in less than two years.

The construction industry has a far-reaching economic role in the United States. With total annual residential and non-residential building expenditures now approaching $400 billion, the industry now accounts for more than five percent of the gross national product, a larger segment than any other industry. Just as significant is the fact that the cost of construction eventually affects the price of all other commodities and services, for businesses must cover the cost of their facilities in their markup for operating expenses. Furthermore, close to one million contractors and subcontractors, employing more than five million workers, make the construction industry one of the largest sources of employment in the country.

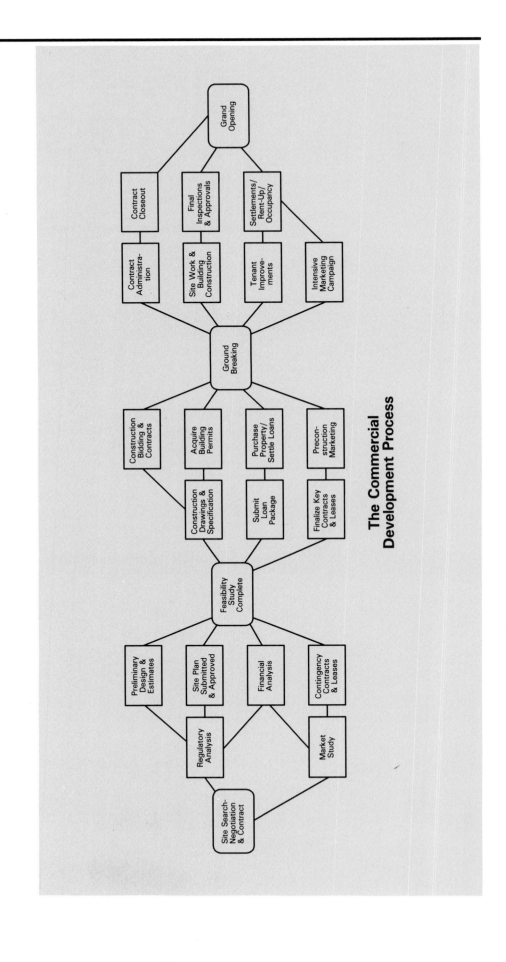

**The Commercial
Development Process**

Figure 1.1

4

Since 1980, several major factors have contributed to the expansion of the non-residential sector of the industry:

1. The Tax Reform Act of 1981 created a wide variety of real estate tax incentives conducive to the construction of commercial and multi-family rental properties.
2. The recession of 1980-1982 resulted in significant pent-up demand for commercial space in many parts of the country.
3. A generally positive economic climate has existed in most regions of the United States since 1983, with the ready availability of financing at favorable interest rates.
4. Demographic and business trends have resulted in a significant increase in the demand for office, retail, lodging, and warehousing facilities.

The effect of these economic influences has been dramatic. Between 1980 and 1985, investment in new commercial and commercial renovation projects increased by 136%, while total residential investment increased by only 62%. However, the positive commercial building climate has been on the decline since 1986, due largely to the overbuilding of large non-residential projects in many metropolitan areas, and the tax reforms enacted in 1986, which significantly diminished the tax incentives for rental properties.

In addition to these negative economic factors, a host of social, environmental, and regulatory issues at the state and federal levels are imposing ever more limitations on the building industry. Many such issues reflect the declining public approval of real estate development in general, and a call for greater social responsibility on the part of builders and developers. While all such issues must be properly addressed, the lack of consistent or clear standards and requirements often results in significantly higher construction costs and time delays.

Despite the many obstacles facing developers and contractors, opportunities exist in most regions for building projects that are smaller, more widely dispersed geographically, and financially structured for income rather than tax benefits. The real challenge is not only in recognizing the best opportunities, but developing the higher level of managerial sophistication required to capitalize on them. It is no longer sufficient to "do it the way we have always done it." The conventional functions of planning, organizing, and controlling must be recast in response to the cyclical and unpredictable business environment facing all businesses today. Success now requires an informed and up-to-date perspective, lean and efficient staffs, creative marketing, and effective human relations skills. A key factor in attaining managerial excellence is understanding and implementing an effective financial control system—the subject of this book.

Characteristics Unique to the Industry

Accounting and other related financial control disciplines are the most neglected control functions in contracting firms. The result of this neglect is a wide variety of operational problems. Often the firm is understaffed and places the greatest value on field production with little attention given to administration. The owners of many firms have little financial background, and are slow to recognize the need for a structured financial control system. In addition, the construction industry's product is unique in terms of the variety, multiple locations, and large contribution of each individual project to the firm's total annual volume. Financial control systems must therefore be designed to accommodate the following special characteristics of the industry.

Bonding

Contractors bidding on a private or public project may be required to post a *bid security bond* or cash deposit equal to a percentage of the total cost. This requirement provides assurance that only qualified contractors submit bids, for the surety company will do a thorough evaluation of the firm's capability in performing a given project. The bid bond guarantees the contractor will either sign a contract for the bid amount, or the surety will pay the difference between the contractor's bid and the bid of the next lowest responsible bidder. The surety then has the right to recover its cost from the contractor's assets.

In addition to the bid bond, a contractor may be required to post a *performance bond*, also available through sureties. This bond provides the owner with protection against the contractor's failure to complete the project. A *payment bond* may also be called for in the contract documents, ensuring that the contractor's vendors and in-house labor are all properly compensated. Again, in the event of a financial loss, the surety has recourse against the contractor's assets.

Because of the risk assumed by sureties, they must investigate virtually all aspects of a contractor's history, financial strength, and ability to perform. Their financial review is typically quite exhaustive, involving not only the standard financial statements, but extending to additional managerial accounting reports. Examples might include job status reports for all work-in-process, cash flow projections, financial ratios and trends, and budget forecasts.

Financing

The cost, form, and availability of financing for contractors is a function of the risks they incur. The greatest need for financing is often *working capital*, because of the lag between expenditures and cash receipts. Commercial banks will often grant a working capital line of credit for a short term, tied to specific contracts. The creditor may accept an assignment of the contract and the related receivables, as well as requiring the contractor to sign personally as security. Many contractors also establish a traditional line of credit for this purpose.

Progress billings on work-in-process can provide a significant source of financing if controlled properly. This method requires a favorable schedule of values (front-loaded for initial phases), efficient scheduling in the field, timely issuance of draw requests, effective follow-up on receivables, and clear contractual terms related to payments. Ideally, the contractor may receive draw payments before many of the expenditures related to the draw are made, providing a short term source of capital.

Decentralized Production

Each construction project requires a new temporary "factory" in the field. Office facilities, field staff, equipment, and administrative support must be available at each site. From the financial control standpoint, decisions must be made at an early stage as to the process for collecting and processing accounting data generated at the project level. The field personnel who typically collect this data tend to be production-oriented and do not appreciate the importance of accounting data or related procedures. Consequently, such data is often delivered late, incomplete, inaccurate, or sometimes not at all.

Managers must establish specific procedures, with appropriate checks and balances, whereby field-related accounting data is collected and delivered to the accounting office. These procedures should include time sheets, contractual data, purchasing information, change orders, and all other field functions that result in financial obligations. On some larger projects, it may be prudent to set up an accounting office of limited scope on-site. However, most contractors find a centralized accounting function within the main office to be the most efficient arrangement.

Dependence on Outside Professional Services

Figure 1.1 illustrates the complexities encountered in planning and executing commercial building projects. Throughout this process, the contractor must rely on a wide variety of specialized outside professionals in order to bring about a successful project. Finding qualified services and coordinating their activities is often the most difficult and time consuming task facing the contractor.

During the planning stage, specialists in market analysis, financial analysis, site planning, and architectural design must be engaged to determine the feasibility of any given project. Following the feasibility stage, a number of individuals skilled in drafting, estimating, government approvals, financing, and marketing become a part of the development team. Some of these functions may be provided internally by the firm, but most often the contractor will depend to a large degree on outside firms or individuals.

During the construction phase, the firm must rely on a wide variety of subcontractors and suppliers, each of whom specializes in a particular phase of the construction process. The firm must implement a specific qualification process; detail-oriented contract preparations must be carried out; and accurate scheduling procedures must be established. These are requisites for a successful relationship with subcontractors and suppliers.

In addition to retained professional services, the contractor also depends on financial institutions, bonding companies, and public agencies. These entities place various demands on the firm's accounting practices, and often require financial reports above and beyond those required in most other industries. The firm must be prepared to present its financial position in the most favorable light and in the format required by each of these groups.

Types of Projects

In contrast to firms that manufacture a large quantity of only one or a few products, contractors produce a small quantity of very different products. Buildings used for such diverse applications as retail sales, offices, warehousing, and manufacturing facilities are all quite different in terms of size, configuration, and construction. Furthermore, each project within these functional categories is typically unique, requiring specialized materials, building techniques, mechanical and electrical systems.

The diversity of construction products and methods results in a need for *job cost* accounting, rather than the *process cost* accounting found in most manufacturing operations. Job cost accounting requires that all direct field costs be traced to a specific project, and that indirect field costs be either directly applied to projects when easily traced, or allocated to a number of projects when not easily traced. Unfortunately, collecting and tracking cost data across any number of ongoing projects is not an easy task, and creates several unusual financial control problems.

To make matters a little more difficult, the job cost function must take place in a timely fashion so that managers can effectively respond to the information. With all costs summarized and reported on a monthly basis, managers are in a position to recognize unfavorable variances and take action before these situations blossom into serious problems. The job cost system also provides historical cost data which can be used when considering projects of a similar nature.

Contractual Obligations

Work performed by contractors, regardless of the type of contractor or project, is governed by an agreement with a customer to build or improve a piece of real property. The contract specifies the work to be performed, the amount and timing of payments, and a number of specific responsibilities borne by both the customer and contractor.

A number of different forms of agreement are found within the industry. A fixed-price contract provides for a single price for the total amount of work to be provided. A unit price contract requires the contractor to perform a specific project at a fixed price per unit of output. Another common approach is a cost-plus contract, in which the contractor is compensated on the basis of all direct construction costs plus a percentage of the direct cost or a fixed sum to cover operating expenses and profit.

Because of the physical size, seasonal weather conditions, and/or design complexities involved with most building projects, up to a year or more is often required for completion. Given the extensive advance time required to prepare drawings and receive permits, the contractor is sometimes obligated to a fixed price for an extended period of time. Action must be taken, through contractual terms, to protect the firm in its relationship with both the owner and vendors.

Those contracts that overlap the firm's fiscal year end (regardless of whether they actually require 12 months to complete) may be treated as long term contracts for purposes of recognizing revenue. The percentage-of-completion and completed contract are alternatives for financial and tax reporting. These methods are available almost exclusively to the construction industry because of the unique character of its products. Chapter 8 addresses these issues in detail.

Risks

Contractors take on many more operational and financial risks than the average business operator. Already mentioned are the lengthy contracts with long term financial obligations. In this era of unstable economic conditions, largely due to the unpredictable level of inflation and interest rates, most people would think twice before entering into such binding, long term agreements.

Accurate Estimates: Although every attempt is generally made to develop accurate estimates, it is practically impossible to ensure total accuracy in every material or labor takeoff. Often construction plans and specifications are not perfectly clear, and some level of assumption is required. Given the high cost of materials and labor, the potential for losses due to inaccurate estimating, even if only slightly inaccurate, is quite high. Therefore, it is crucial to the financial stability of the firm that specific procedures—both in estimating and subsequent purchasing—be implemented and monitored to limit this vulnerability.

On-site Safety: On-site safety is a major liability assumed by contractors. There is a high likelihood of accidents in construction, and every effort must be made to safeguard against unsafe conditions that may bring about an accident. The Occupational Safety and Health Act

(OSHA) outlines an extensive list of regulations and standards to be followed in this regard. Contractors should carry comprehensive insurance coverage to protect the assets of the firm in the event of a lawsuit resulting from a serious accident.

Avoiding Legal Actions: Given the large financial investments involved in nearly every step of the development process, as well as the complexities inherent in the actual construction process, the potential is great for conflicts between parties. Almost every day, newspaper articles appear describing lawsuits between contractors, owners, designers, and subcontractors. The contractor must protect himself against these risks by constantly monitoring his own operations and carefully preparing each contractual agreement before entering into working relationships with other parties. Professional guidance from attorneys experienced in construction law is crucial. Again, appropriate insurance coverage is essential.

Customer Relations

Contractors assume any one of several roles in their relationships with customers. Traditionally, they have not participated as members of the development team until late in the game, simply providing a bid on contract documents prepared by the owner's architect or engineer. This is an "arm's length" transaction in which the contractor assumes only the responsibility for completing the work in accordance with documents provided by others and within the accepted bid. In many cases, the owner's agents continue to represent him throughout the project, isolating the contractor from his own customer.

Over time, this traditional approach has proven unsatisfactory for many owners. They must engage the services of multiple design firms, coordinate their activities, and pay for all services rendered prior to confirmation of the total construction costs involved. Too often, complete contract documents have been prepared and then scrapped because the cost was excessively over budget. The time alone that is required to go through this exercise is often unmanageable in today's rapidly changing business environment.

New relationships have been developed between contractors and owners in the form of design/build and construction management services. In both cases, the contractor becomes a member of the development team very early in the process, and assumes somewhat of an agency role in representing the owner. The financial arrangements are often different from the lump sum commitment made in the competitive bidding approach.

Trends in Construction

A number of identifiable trends are affecting the construction industry today. Managers who recognize and adapt to these trends will have much better prospects for both survival and superior profits in the future. Interestingly, many of these trends are the direct result of the recognized need for greater financial control.

Project Size and Location

Thousands of large commercial projects, driven largely by tax incentives, have been built in many U.S. cities since the early 1980's. There is now an oversupply of virtually all types of commercial space in many areas, with extremely high vacancy rates. It is expected there will be a move in the future toward smaller and more geographically dispersed projects, creating excellent opportunities for small and medium sized contractors seeking diversification opportunities.

Automation

Contractors, large and small, have turned to computers for the competitive and managerial edge required to earn superior profits. Among the functions that are automated in many advanced contracting offices today are accounting, cash flow forecasting, estimating, scheduling, and drafting. The *financial control* offered by computers is perhaps the most significant benefit, providing more accurate and comprehensive information on a much more timely basis.

Design/Build

The traditional relationship among architects, owners, and contractors has changed dramatically in the light commercial market. The design/build and construction management approaches offer a number of advantages in terms of time and cost savings. Those contractors seeking a secure and profitable future would be well served by investigating the features of these services and pursuing those clients most likely to use these alternatives to the traditional competitive bid approach.

Competition

Competition in the contracting industry is at an all time high, with profit margins at an uncomfortably slim level. With more contractors available to do fewer projects, those seeking superior profits must seek out their market niche and operate as efficiently as possible.

Specialization

As a direct result of the competitive business climate, many firms have found the generalist approach to services is no longer the ticket to profits. Today, the smaller, faster, and more specialized firm that can most efficiently address the needs of a specific client on a specific project will find the profitable niche.

Marketing

Many contractors are content to simply ensure their presence on a large number of bidders lists, and to "win" their share of the market through the statistical game of competitive bidding. However, most successful contractors have discovered the value of planning and implementing an ongoing marketing effort. It is no longer enough to just have plenty of marginally profitable work. Aggressive managers are using effective marketing techniques to make superior profits on those specific projects most suited to their firm's abilities in terms of project size, type of construction, type of customer, and geographical location.

Decentralization

A natural outgrowth of the complex business environment is a move by medium to large contractors toward decentralized control. Many of the responsibilities previously assumed at the executive level are now delegated to middle managers, and responsibilities previously assumed by office managers have been delegated to field managers. In most cases, staffs have been trimmed, placing additional demands on reduced numbers of workers.

Scarcity of Tradesmen

Many commercial building techniques and materials require specialized skills, yet competent tradesmen are becoming more difficult to find and retain. Economic recessions result in massive industry layoffs, causing many workers to seek more stable employment. Improved financial control systems adopted by construction employers allow more stability in staffing, as well as the compensation packages most desirable to qualified tradesmen.

Subcontracting

Up until the last 10 to 15 years, many contractors maintained their own crews and equipment to perform basic trade work such as carpentry, concrete, and masonry. This approach seemed to allow the greatest control in terms of cost, scheduling, and quality. While having one's own crew still offers some distinct advantages, the tremendous investment in equipment and personnel has made this alternative less practical in today's business environment. As the economy takes its swings, contractors must have the flexibility offered by subcontracting to expand and contract quickly, without incurring the long term debt and expansion of overhead expenses associated with field crews.

Renovation

Renovation is now the fastest growing sector of the commercial market, and is quickly approaching new commercial construction in terms of total dollar investment. The multiple advantages of rehabilitating older structures in downtown areas will continue to capture the attention of owners and developers, particularly as construction and land costs rise and the selection of new locations diminishes.

Investment Strategies

The substance and utility of the tax incentives which have driven commercial real estate in the past have been severely limited by recent tax reforms. New rules for depreciation, active vs. passive income activities, capital gains, and tax credits are forcing developers and owners to approach their projects from an income and appreciation angle in lieu of tax benefits. Contractors who understand these changes can better advise their design/build clients as to the most viable financial structure for their projects. Many contractors also take a lead role in packaging development projects and then selling partnership interests to raise the equity capital needed.

Government Regulation

The level of government regulation in the building industry is rapidly escalating and touches virtually every aspect of a contractor's operations. Many firms have resorted to hiring a full-time staff to accomplish what used to be such relatively simple tasks as acquiring site plan approvals and building permits. Those firms engaged in projects financed with government funds quickly find a whole new world of special requirements. Given the growth in environmentalism, consumerism, and bureaucracy in general, this trend will only escalate in the future.

Managerial Sophistication

In light of all the other trends affecting the commercial construction industry, contractors and developers alike have found the need to establish more sophisticated managerial and financial control systems. More college-educated managers are adopting the most up to date planning, organizational, and control techniques that are now so crucial to the efficient operation of their businesses.

Why Contractors Fail

The managers of new and existing construction firms can get an inexpensive but valuable education by studying the causes of the high business failure rate in the construction industry. By understanding these causes, managers can avoid making the same mistakes and dramatically improve their own prospects for success. Poor financial control plays a leading role in many of the causes of failure. While it may show up initially in terms of low profitability, its roots may be found in low productivity, poor cost control, inadequate cash flow, and generally poor managerial control.

Lack of Knowledge

The commercial construction industry is highly complex and involves a very broad range of business and technical disciplines. In addition to basic construction practices, other disciplines include architecture, engineering, accounting, finance, marketing, law, and zoning. The owners of many building firms start with a high level of expertise in one or two of these disciplines, but fall short of developing the minimum working knowledge required in all associated fields.

Given the highly competitive nature of the industry, only those contractors who take the time to understand the "big picture" and obtain professional guidance in all specialty areas will both survive and remain profitable. The truly smart contractor is one who knows what he *does not know*, and is not too proud to ask for help. In fact, most successful managers will not make significant business decisions without first consulting with their professional advisors.

Unrealistic Prices

Profitability in any business endeavor requires total revenues that exceed total expenditures for direct costs and operating expenses. Unfortunately, many contractors simply underprice their services, ensuring their own demise no matter how intense or efficient their subsequent efforts. Amazingly, some managers apply unit prices or overhead markups based on a historical rule of thumb heard years earlier.

Avoiding this very basic error requires thorough and accurate estimating of all the direct costs associated with completing a given project, application of appropriate markups to cover all costs of doing business, and finally, applying a markup to provide a reasonable profit. These markups must be developed through a budgeting process that anticipates the percentages of fixed, variable, and semi-variable expenses that exist at various levels of volume and gross margins.

Competition

Excessive competition may sometimes prevent contractors from securing work with adequate profit margins. However, more often than not, the contractor who fails to *recognize* and *respond* to the competition is the one who will encounter severe problems. Some contractors try to be all things to all people, thereby diluting their effectiveness and ability to compete. Others unknowingly offer the wrong services in the wrong markets.

Successful contractors first analyze their competition in terms of their strong and weak points. There is no need to go "head to head" against a firm with more expertise and a better competitive position on a specific type of project. Managers are better advised to seek out a market niche that is not thoroughly addressed by the competition and target their marketing efforts in such a way that their firm becomes the leader in that specialty area.

Inadequate Capitalization

Practically all building firms take on more work than they can comfortably handle at one time or another. The overload is often manifested in poor quality, field delays, and angry customers. However, the most financially damaging outcome may be a shortage of working capital to cover the direct construction costs and operating expenses incurred by the firm. As payments to subcontractors and suppliers become overdue, discounts are lost, credit terms are restricted, and relationships are strained. In an effort to preserve cash, the firm may cut back on such important items as marketing and wages, which in turn cripples the firm's ability to obtain future projects and retain skilled workers.

Bonding firms, well known for their conservatism, typically require contractors to carry total work-in-process that equals no more than ten times working capital. Therefore, a firm with $200,000 in working capital should theoretically attempt no more than $2 million dollars in work-in-process at any given time. Since this standard is often exceeded by many contractors, the level of financial control throughout the organization then becomes an even more significant issue in avoiding a cash shortfall.

Slow Collections

Given the "ripple effect" of inadequate capitalization, contractors must carefully monitor their cash position at all times, not only in terms of workload, but also collections. While they are at times justified in delaying partial or full payments, some owners may simply use contractors as their personal "banker", by borrowing large sums in the form of withheld payments, without authorization and without interest. All too often, contractual agreements do not sufficiently protect the contractor from customers who "drag their feet" when it comes to satisfying draw requests.

The burden is upon the contractor to ensure that the proper language is included in the contract, clearly establishing the sums to be paid, when they are to be paid, and the penalties to be applied for slow payment. Just as importantly, the contractor must establish specific collections procedures to monitor and follow up on all receivables before they become excessively overdue. Inadequacies in field operations that might cause owners to withhold payments must also be avoided.

Lack of Cost Control

One might think a contractor is guaranteed a profit if he implements accurate estimating techniques and applies realistic markups. However, what starts out as a profitable project may soon turn sour because of inefficient cost control in the field. This inefficiency may result from overpaying for materials, wasting materials, insufficient negotiations with subcontractors, a poor system for monitoring job costs, low productivity of workers, or a variety of other shortfalls.

Purchasing policies and financial systems must be established and continually enforced to maintain control. Bidding of all subcontract phases and preparation of tight agreements by an experienced contracts administrator is a crucial factor. A job cost system for tracking and reporting on cost variances must be implemented and monitored by top management on a routine basis. Finally, an effective financial reporting system must be incorporated that provides current and properly formatted data useful to management in planning and decision making.

Insufficient Quality Control

Poor quality in the field strikes the general contractor first by driving up the costs for correcting deficiencies, and later by having to execute warranty repairs. Warranty expenses can easily become a constant and unpredictable drain on the bottom line. As quality goes, so goes the firm's reputation and long term prospects for survival. Given the trend towards design/build and construction management services, continuing problems with quality control will quickly catch up with an otherwise well managed and profitable firm.

Management must establish and monitor an ongoing quality control program that includes specific procedures to be followed. A system of automatic checks and balances must be designed to ensure all standards are met before payments are made to subcontractors. Field managers should receive all the required technical training to ensure their ability to recognize and control quality.

Inadequate Marketing

Contractors are notoriously poor in marketing their construction services and developing the very image that will bring superior profits. Amazingly, many contractors make no attempt whatsoever at even a minimal marketing program, being satisfied to simply "knock heads" in one competitive bid situation after another. Others seem content so long as they have plenty of work, regardless of the profitability of that work.

The more successful firms are allocating a greater portion of their budget than ever to marketing their services. The goal is the image building and target marketing that will help to secure a profitable market niche and reach the most likely prospects. Most often, this effort must include the guidance of a marketing professional who can assist in developing the marketing plan and implement specific strategies.

Poor Human Resource Management

How many times have we all heard, "You just can't get good help anymore?" Unfortunately, it *is* very difficult to find skilled field and office personnel, and even more difficult to retain them. Subcontractors and tradesmen today are more independent than ever, with many options available to them. Many managers have simply not learned to manage people on a level considered positive and productive by today's standards. The result is either rapid turnover of personnel and subcontractors, or low productivity, which, in turn, causes delays, confusion, and higher costs.

In all cases, a contractor's survival depends on successfully developing and managing people who respond more favorably to positive than to negative inducements. A pleasant and safe working environment, productive relationships among workers, positive reinforcement, and a well-rounded compensation plan are all major factors in attracting and retaining qualified and motivated personnel.

Lack of Business Planning

Many businesses of all types operate with no written business plan, or worse, no business planning function. Managers complain about the lack of time to do business planning, or believe it is not practical in such a volatile environment. Most often, managers find themselves so caught up in a day-to-day effort to "put out fires", that they never step back to plan and control the "big picture."

Unfortunately, an organization without a business plan is like a ship in dark and stormy seas, without navigational aids or the ability to communicate. The ship is unlikely to reach its destination under these conditions and is in danger of sinking or ending up in a far different location than was intended.

Profitable contractors must take the quality time needed to develop specific objectives important to the success of the firm. Specific areas of concern typically include *sales volume*, *marketing efforts*, *changes in organizational structure*, and a number of *financial goals*. Strategies capable of accomplishing desired objectives must be defined and communicated to those individuals responsible for producing the desired results. A control system must also be implemented to monitor the success of the firm in accomplishing its objectives.

Evolution of Financial Controls in the Firm

The financial control function changes dramatically as a construction firm evolves from a small operation into a substantial multi-million dollar enterprise. At the early stages of development, when a single proprietor provides practically all managerial functions, financial control is likely to be limited to maintaining the general ledger and a minimal payroll.

As the firm grows, accounting and other financial control systems become increasingly important. As shown in Figure 1.2, most contractors go through three identifiable stages as their firms mature. We will call these stages the *Upstart*, the *Veteran*, and the *Strike Force*. These stages do not necessarily coincide with growth in sales volume, for many firms intentionally limit their growth while focusing on product mix and profitability. While others may label these stages with more traditional terminology, the progression from one stage to the next is almost always the same.

At the Upstart Stage of developing a construction firm, most contractors focus on the area in which they are most comfortable— field production. The philosophy is that if the projects are built correctly and quickly, profitability will most certainly follow. However, a number of other crucial administrative functions are often set aside during this formative period.

The Veteran has survived a few battles during the formative years, but has not won the war. The contractor at this stage realizes that far more is involved in earning profits than simply constructing buildings. It is at this point that the contractor begins to develop a better understanding of the external issues affecting his operation. The need for more attention to planning and marketing also becomes apparent. Attempts are made to develop policies and procedures, to improve financial control systems, and to better define organizational structure. These are all very worthy aspirations. Unfortunately, successfully attaining these goals requires an investment of time and a managerial expertise that many contractors simply do not possess. As a result, many construction operators never proceed beyond the Veteran stage, accepting mediocre growth and profits while never recognizing the potential that could be realized through more intensive managerial and financial control.

The contractor has arrived when he reaches Stage 3—the Strike Force. At this point, he has probably recognized his own limitations, and sought out the expertise of qualified external associates (lawyers, accountants, marketing experts). A whole new appreciation for managerial and financial control develops, and extensive efforts are made to bring the firm up to the level of sophistication required to earn superior profits.

Stage	Managerial Control	Financial Control
1 The Upstart	Little understanding of business environment No business plan—unspecified goals Poor marketing effort Minimal attention to organizational structure Centralized control—generalized tasks Unsystematic, inefficient, unproductive Good field knowledge, average field control Poor office control—manual systems No dependence on external specialists	Minimal appreciation for financial control Inappropriate accounting system No tax planning No budgeting—improper markups Bookkeeping internal—acctg. by external service Inadequate job cost control Poor asset management Annual or quarterly financial accounting reports No managerial accounting reports
2 The Veteran	Partially responsive to business environment Informal business planning—general goals Improved but incomplete marketing effort Preliminary attempts at organizational structure Partially decentralized control Systems developed—improved efficiency Good field knowledge—good field control Average office control—partially automated Limited dependence on external specialists	Need for good financial control recognized Improved accounting system Informal tax planning Informal budgeting—better markup analysis Internal accounting—partially automated Attention to job cost control Attention to asset management Quarterly financial accounting reports Selected managerial accounting reports
3 The Strike Force	Fully responsive to business environment Formal business planning—specific goals Formal and effective marketing plan Formal organizational structure Decentralized control—specialized tasks Very systematic and efficient Good field knowledge—excellent field control Excellent office control—fully automated Substantial dependence on external associates	Sophisticated financial control system Refined accounting system Formal tax planning Formal budgeting—markups Fully automated internal accounting Sophisticated job cost control Sophisticated asset management Monthly financial accounting reports Extensive managerial accounting reports

Figure 1.2

Planning and developing strategies become institutionalized, and every aspect of the firm's financial and organizational health is regularly diagnosed and improved. At this stage, the contractor is seeking the ultimate in efficiency and profitability.

The financial control system found at the Strike Force stage has become very refined, starting with a tightly controlled internal accounting system. Firms at this stage of development routinely engage tax specialists to analyze and minimize the tax exposure of the firm, as well as to direct the firm in major asset management decisions. High quality historical data is used and careful forecasting is performed in order to produce annual budgets. This research and forecasting process provides "accurate" numbers to be applied as markups on direct costs. Meaningful financial and managerial accounting reports are provided to upper management at regular intervals, allowing rapid response to problem areas. Well defined policies and procedures are adopted to direct all affected staff members in their responsibilities related to the financial control system.

Figure 1.3 illustrates the organizational changes that occur in a light commercial contracting firm experiencing increasing sales revenues. The actual structure of a given firm is likely to differ from those shown in Figure 1.3, for a number of internal and environmental factors will influence organizations in varying ways. However, Figure 1.3 does reflect the kinds of changes that may be reasonably anticipated as growth occurs. Most important to this discussion is the evolution from the secretary with partial bookkeeping responsibilities at the early stages, to the full-blown finance department that may be required when the firm reaches a sales volume of $20 million.

Growth of the Finance Department

As the firm experiences successive growth stages, the role of the "chief number cruncher" changes dramatically. At the early stages, this individual needs only limited skills to perform primarily bookkeeping functions. As the accounting system becomes more demanding, an accountant may be employed who is capable of coordinating the entire system, including the preparation of financial and managerial accounting reports. As the firm reaches yet higher levels of revenue, someone with a great deal of financial management experience may assume the position of Controller, or Vice President of Finance. This individual is responsible for virtually all of the financial affairs of the firm.

Organization

Position descriptions are an important internal control document for all firms, outlining the specific duties, organizational relationships, and authority associated with any given position. Each employee within the firm should be provided with the appropriate written description of his or her position, and instructed in the details of what is expected. As the firm grows, these position descriptions must be updated to reflect the changes that naturally occur as individuals take on expanded duties. Such descriptions should reflect the responsibilities, relationships, and authority that may be assigned to individuals in these positions.

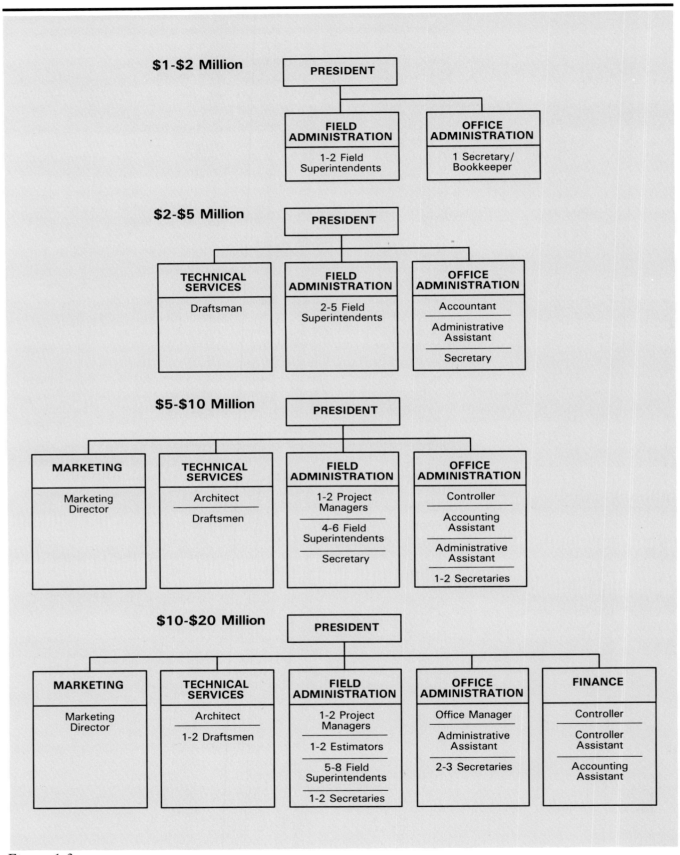

Figure 1.3

Each firm must evaluate its own specific circumstances in preparing such position descriptions. Organizational structure, sales volume, work volume, and other factors create widely diverse demands on the same positions within different companies, and within the same company at different stages of growth. For example, as the firm reaches higher levels of volume, the accounting function may become more departmentalized, with certain individuals assigned to payroll, others to job costing, and yet others to payables and receivables.

CPA Services

In addition to the internal accounting functions outlined in the sample position descriptions, most firms must also engage the services of a certified public accountant (CPA) to assist in the preparation of year-end financial statements. A CPA would also be called upon to advise the firm on complex tax-related issues. If required by creditors, the firm's CPA may audit the firm's financial statements through an in-depth evaluation of all financial data and a verification of its accuracy. Most public accounting firms also provide Management Advisory Services (MAS), such as the design of financial control systems, forecasting, financial analysis, and computer system installations. Given the unique operational and accounting characteristics of the construction industry, contractors should seek out a CPA or business consultant with specific construction-related experience.

Financial Control Disciplines

Many contractors consider financial control as primarily a bookkeeping function, the purpose of which is to simply keep the records straight and produce a year-end financial statement. Indeed, such routine tasks as payroll, accounts payable, and accounts receivable are largely number-gathering and organizing exercises. However, practical applications involving a contractor's financial data go far beyond bookkeeping and basic financial statements. Such data is the source of managerial accounting information, used by contractors in planning and controlling the firm's financial activities.

The financial control function, as shown in Figure 1.4, takes in a number of disciplines, all of which interact in terms of data accumulation and corporate goals. There are few hard and fast rules regarding these disciplines, for each firm operates in its own unique way depending on its size, products, customers, location, and goals. Therefore, specific financial control systems must be evaluated and established for each firm and its methods of operation.

A brief overview of these disciplines follows, with more in-depth analysis provided in subsequent chapters.

Construction Accounting

Accounting is the process of collecting, summarizing, and reporting financial information concerning a business entity. The methods used by a business to record and summarize its financial activities is referred to as its *accounting system*. This system is used to record the monetary transactions made by the firm, such as purchasing materials, borrowing funds, paying subcontractors, and receiving payments from owners. Each of these transactions must be recorded and classified into related groups. These groups and classifications are then summarized into accounting reports that provide valuable information to the firm's managers and external associates.

Financial statements are the final output of the financial accounting system, and are used to summarize the transactions of the firm for a given period of time. The two most widely used financial statements are the *balance sheet* and the *income statement*. The balance sheet reflects the financial position of the firm at a particular date in terms of assets, liabilities, and net worth. The income statement shows the total revenues and expenses for a given period, with the difference representing profit to the firm.

The information contained in these financial statements, as well as other reports generated by the accounting system, is the source of data used in the other financial control functions. For instance, *historical income statement* data is used extensively in preparing operating budgets for

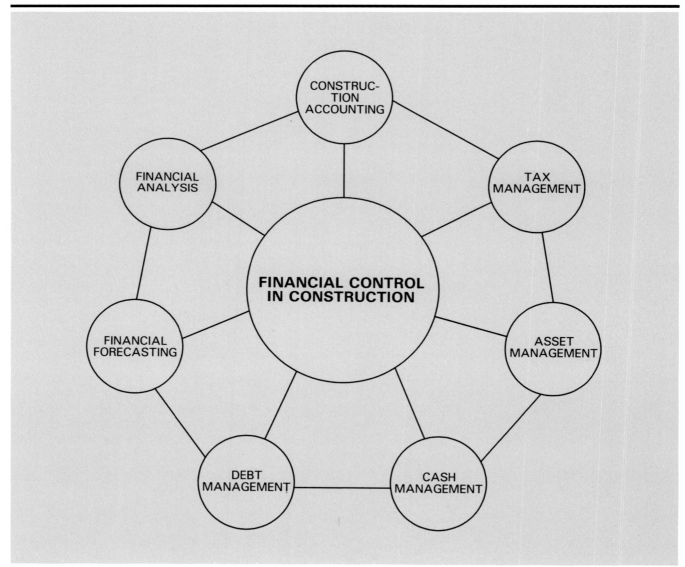

Figure 1.4

future periods. *Current period revenue projections* are used in evaluating asset acquisitions and planning strategies to minimize the tax exposure of the firm. It is easy to recognize the value of these reports to management for forecasting future conditions and developing specific goals.

Financial Analysis

A number of procedures have been developed to assist managers and creditors in evaluating the financial statements and associated performance of businesses. The strong and weak characteristics of a firm are exposed through financial analysis which, in turn, allows the planning and restructuring required to yield improved performance. The two most common procedures in this regard are *trend* and *ratio analysis*.

Trend analysis simply tracks the changes in specific financial data over a number of accounting periods. For instance, income statement data may be consolidated for a three-year period, displaying in "snapshot" form the changes in revenues, direct costs, operating expenses, and profits during this time. Both positive and negative developments in any of these categories allow management to understand how and why the firm has become more or less profitable.

A financial ratio is simply a mathematical expression of the relationship of one piece of financial data to another. A ratio focuses attention on a significant relationship, and provides valuable insight into the financial health of the firm. A common ratio is the *percentage of profitability*, derived by dividing profit by revenues. For example, if net profit before taxes in a given period is $80,000, with revenues of $2,000,000, the percentage of profit is $80,000 divided by $2,000,000, or 4%. This percentage can then be compared with that of prior periods as well as with industry standards for profitability. A wide variety of such ratios has been developed to analyze both balance sheet and income statement data, as well as non-monetary data.

Trend and ratio analysis is used by managers to evaluate their firm's current condition, and to plan the course of future activities. The goals of individual firms vary, but almost all of them want to improve both profitability and stability. Many alternatives are available to managers as they seek these goals through effective financial planning.

Financial Forecasting

Nearly all organizations prepare an annual budget, expressing the sales volume and operating expenses anticipated during a specified period, typically one year. The budget is a *profit plan*. It reflects the results of activities planned by the firm in an effort to attain its profit goal.

The contractor's budget consists of several components. The first is anticipated *sales volume*. Sales volume is a composite of the volume from current projects, and projects that the firm expects to acquire. This projection is an output of the marketing plan, and is predicated on a number of forecasting assumptions.

The second major component is *expenses*, which are organized into three categories. *Direct costs* cover the actual labor, material, subcontract, and equipment costs incurred directly through field production. *Indirect costs* are those incurred in the field, but not easily attributed to one specific project. Examples are field-related insurance, small tools, and company benefits. Indirect costs are typically allocated among all projects through percentage markups. The last category, *general and administration expenses*, covers general operating expenses that are not field-related. Examples include office salaries, rent and utilities, and marketing.

The last component of the budget is *profit*, the difference between anticipated revenues and expenses. Once an accurate budget is prepared, the revenue and expense projections may be manipulated in response to various alternatives. These alternatives might include increasing the marketing effort to pump up revenues, decreasing operating expense, tightening up on direct costs, or increasing markups. Most importantly, the budget provides a plan, or goal, that managers can strive to accomplish through effective financial control systems. It is often desirable to establish and monitor budgets on a departmental basis. In this way, the performance of individual departments or managers can also be monitored.

Asset Management

Contractors employ a number of assets in their day-to-day operations. Planning the acquisition of these assets and ensuring they are effectively utilized is a key to the profitable operation of the firm.

Assets can be classified as current or capital. Current assets are those that will be converted to cash or used up within one year and include cash, accounts receivable, and inventory. Capital assets provide the firm with benefits for periods beyond a year, including trucks, equipment, and office facilities.

Unfortunately, many contractors make the mistake of "loading up" with capital assets. The long term debt associated with these assets can be overwhelming in poor economic conditions, and must therefore be carefully considered. A number of methods are available to help managers make decisions as to the purchase versus leasing of such capital assets.

Contractors often incur major cash flow problems due to a lack of control over their receivables. Substantial delays can cause immediate difficulty in meeting payables, in turn causing a poor credit rating and conflicts with creditors. Effective contractual protection is crucial to ensuring receivables control. Procedures must also be established for follow-up in collections.

Debt Management

Contractors are heavily dependent on debt due to the large dollar value of the projects they build. This debt may be short or long term, depending on the length of time it is borrowed and for what purpose. Short term debt must be repaid within one year or less and is typically used to finance immediate working capital requirements or short term expansion needs. Long term debt is arranged for periods exceeding one year. Generally it is used to acquire capital assets, which have a life exceeding one year.

A number of issues must be considered in negotiating short or long term financing with creditors. Among them are the term of the loan, interest rate, repayment terms, and collateral. Managers must have a working knowledge of the many alternatives available in order to minimize their exposure and expense when borrowing funds.

Cash Management

Given the large dollar value of their projects, contractors are faced with a constant movement of substantial quantities of cash in and out of their firms. Understanding and managing this movement is crucial to the stability of any contracting firm, for a failure to do so has led to the bankruptcy of many firms that were otherwise well managed. Cash management involves not only planning and controlling cash flow, but also the effective utilization of cash as an income-producing asset.

A number of cash management techniques are important to the profitability of any construction firm, including:

- Budgeting the firm's cash needs
- Accelerating cash receipts
- Decelerating cash disbursements
- Raising cash at the lowest possible cost
- Investing excess cash at the highest possible return

Contractors will realize many benefits from properly observing these techniques. In addition to the obvious advantage of maximum liquidity, other important benefits include higher profits and improved bond limits. Managers who are able to display a high degree of managerial control in all aspects of their operations, including cash management, will be viewed much more favorably by all of their creditors.

Tax Management

Our silent partner, the Internal Revenue Service, expects a share of all profits generated by business. Successful tax management allows the firm to retain a greater share of its profits, as well as the income it can receive by investing those funds. The challenge to managers and their advisors is to defer taxes as long as possible, and avoid taxes altogether when feasible. Under ideal circumstances, taxes may be deferred from one period to the next, allowing the firm to "borrow" funds tax-free from the government.

Tax planning should not be an annual, last minute effort to reduce the firm's tax liability. It should instead be an ongoing effort, so that the firm's tax obligation is not a surprise that places an undue strain on cash reserves. The firm's tax exposure should also be a definite factor when important decisions are made concerning the acquisition of capital assets.

The Tax Reform Act of 1986 dramatically altered the tax structure affecting all businesses in this country. Many rules were changed and those changes had a direct impact on contractors. The following areas were particularly affected.

- Organizational Structures
- Accounting Methods
- Tax Rates
- Capital Asset Ownership
- Fringe Benefits
- Real Estate

More than ever, successful contractors must carefully monitor the tax aspects of their businesses and engage the services of specialists who can guide them through difficult tax-related decisions.

Chapter Two

CONSTRUCTION ACCOUNTING

Chapter Two
CONSTRUCTION ACCOUNTING

The success of any construction firm is closely related to its ability to forecast and control costs. Both of these functions depend on a properly designed and operated accounting system. Unfortunately, construction accounting can be quite complex, and the industry has a history of failing to perform this function properly. Many firms have folded because they have not understood their real costs of doing business.

There are a variety of reasons why contractors tend to neglect the accounting function. Some managers are simply not financially-oriented, and focus instead on project management. Many construction firms are understaffed in the office, and accounting procedures are therefore given secondary attention. However, the industry's product is unique, and requires a system capable of dealing with the nuances found in the construction accounting process.

Financial accounting involves collecting, classifying, and reporting all the financial data and transactions generated through the firm's ongoing activities. The results of these efforts are financial statements, including the *Income Statement*, *Balance Sheets*, and *Statement of Cash Flows*. These statements are prepared not only for management's use in tracking the firm's performance, but also to satisfy the needs of the external parties upon whom the firm must depend for credit. External parties requiring financial data on a construction firm include the following.

- Lenders
- Sureties
- Suppliers
- Public agencies (IRS, Social Security Administration, etc.)
- Clients
- Credit reporting agencies

Too often, accounting is viewed as an exercise in bookkeeping. Financial accounting does, in fact, require the gathering and organizing of many numbers in order to generate a payroll and to track accounts receivable and accounts payable. However, practical applications of financial accounting data are also directly related to numerous other financially oriented management disciplines that are often referred to as *managerial accounting*. These functions are discussed in detail in the following chapters, and include:

- Financial Analysis
- Financial Forecasting
- Cash Management
- Asset Management
- Debt Management
- Tax Management

Construction accounting is a very complex subject, involving far more detail than it is possible to cover in this one chapter. Many excellent books are available that provide this detail. The purpose here is to present general information on how a contractor's accounting system works and what information a manager should receive from it. In addition, we will establish a framework for understanding the topics addressed in the following chapters. For those readers who are unfamiliar with specific terms, a Glossary is provided at the end of the book.

The Construction Accounting Process

An accounting system, for a construction or any other type of business, involves the input, processing and output of financial data on a cyclical basis. Each complete cycle consists of twelve monthly accounting periods, with end-of-period processing at the close of each monthly period, as well as at the end of the yearly cycle.

Elements of the Accounting System

The accounting system consists of *asset*, *liability*, and *equity accounts*. Assets are everything of value held by a firm. Examples are cash, accounts receivable, and equipment. Liabilities are claims against the firm's assets by creditors. Examples are accounts payable, withheld taxes, and notes payable. Equity represents the claims against the firm's assets by its owners. Examples include stock, paid-in capital and retained earnings. Because liabilities and equity represent the total claims against assets, *assets must equal liabilities and equity*. This statement is referred to as the *accounting equation*. The balance sheet, the financial statement that reflects the firm's financial position at a given point in time, is an expanded expression of this equation. All activity of the firm is reflected in this equation, either directly through changes in the assets, liability, and equity or through revenue and expenses, which impact the equity. Net profit or loss is the result of the revenue and expense activity for the current accounting period, which results in an increase or decrease to the equity.

A construction accounting system varies from other accounting systems primarily in the importance of job cost data accumulation, and the corresponding need to carry detailed data forward through several accounting cycles, until project completion. Contractors are more dependent than most businesses on good job cost control in order to determine profitability on each project. Fortunately, direct costs for labor, materials, subcontractors, equipment, and miscellaneous expenses on each project are easily identifiable with an adequate accounting system. A basic understanding of what makes up the job cost input, how it proceeds through the system, and what the output means is essential to effective management of resources.

Chart of Accounts

Accounting records are presented in several different ways, each for a different purpose. The first step in setting up an accounting system is creating a *chart of accounts*. We begin by setting up a separate numerical account for each item that will be part of the financial statements. The classification and order of the items should correspond with the presentation of the statements.

Generally, the chart lists each account with a corresponding account number and account type, with groups of numbers for each category. The divisions must include *assets, liabilities, equity, revenue* and *expenses*. The chart of accounts will be used in the creation of all accounting records and ultimately in creating the financial statements. A chart that can be used to organize and produce the reports described in this chapter is shown in Appendix A.1.

Source Documents

Figure 2.1 is a flow chart of the multiple functions that make up the construction accounting process. The first element of the system is *input*, or *source documents*. For construction accounting, source documents include:

- Time sheets or cards
- Invoices
- Contracts
- Change orders
- Purchase orders
- Estimates
- Draw schedules
- Cash receipts
- Any other document that affects the financial situation of the firm

Any number of various formats can be adopted for these documents. It is, however, important to establish a set of policies and procedures for generating and handling these documents. Having a set of proven procedures provides a sound basis with which to proceed to the second element, the *processing*. A sample set of financial accounting policies and procedures is provided in Appendix A.1.

Journals

Journals are records of original data entered into the system. They may be organized in the form of the general journal and various specialized journals. All transactions are recorded in the applicable journal based on the source document, with reference to the appropriate account from the chart of accounts. The use of these journals facilitates double entry accounting, a method which requires that debits and credits must balance. This means that an entry must be made to at least two accounts for each transaction. Special journals are used to group together the transactions that affect specific accounts. The most commonly used special journals include:

- Cash receipts
- Cash disbursements
- Sales
- Accounts payables or purchases
- Accounts receivable or billings
- Payroll

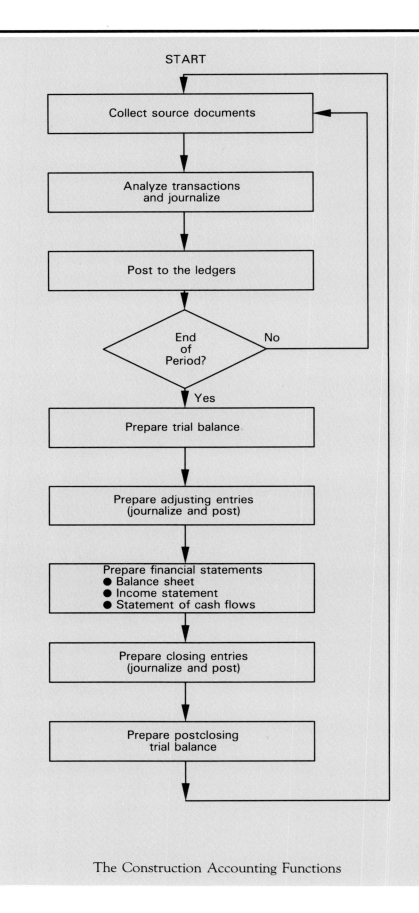

START

Collect source documents

Analyze transactions
and journalize

Post to the ledgers

End
of
Period?

No

Yes

Prepare trial balance

Prepare adjusting entries
(journalize and post)

Prepare financial statements
● Balance sheet
● Income statement
● Statement of cash flows

Prepare closing entries
(journalize and post)

Prepare postclosing
trial balance

The Construction Accounting Functions

Figure 2.1

A sample manual cash receipts journal is shown in Figure 2.2. A general journal is used to record adjusting entries. These adjustments include end of period adjustments such as *depreciation recognition* and *percentage of completion adjustments*. Most computerized construction accounting systems utilize a job cost journal to record transactions initiated within that application, such as correcting transfers of costs from one job to another.

Ledgers

From the journals, transactions are posted to *ledgers*. Ledgers are accumulative records of transactions affecting particular accounts. Each account in the chart of accounts is a *general ledger* account, used to accumulate all data affecting that account. *Subsidiary ledgers* are used to accumulate detailed data for individual components of a ledger. For example, a subsidiary ledger is needed for each vendor within the accounts payable account and for each customer within the accounts receivable account.

Output

The accounting system output, comprised of financial statements and managerial reports, is created from the ledgers. Balances from general account ledgers are used to format a Balance Sheet and Income Statement, while subsidiary ledgers help create management reports such as Job Cost Reports, Accounts Receivable Reports, and Accounts Payable Reports.

Each accounting period, a multitude of tasks must be undertaken to create the output. Figure 2.3 outlines the most significant of these tasks, which can be grouped by type of application, such as payroll, accounts payable, accounts receivable, and job cost. The ultimate objective is access to all of these segments, so that accurate reports can be created on a timely basis for two purposes: making sound business decisions, and satisfying the needs of external parties. In computerized systems, each application is generally assigned to a separate module. The modules are then integrated as shown in Chapter 9, Figure 9.3.

Payroll and Employee Information

Routine payroll processing includes the input in the form of recording time (by employee and job), and calculating wages and all applicable withholdings and accruals for taxes and voluntary deductions. The output is the generation of payroll checks, timely tax deposits, distribution of payroll, and filing of all data.

Maintenance of employee information is extremely important as well, particularly with the ever increasing legislation and liability issues that affect the business environment. At the end of each quarter, additional processing for payroll includes filing Federal, State, and unemployment tax reports, along with W-2's, and 1099's at the end of the year. The importance of timely filings cannot be overemphasized considering the penalties and interest charges for lateness. For example, a late payroll tax deposit penalty is ten percent of the deposit.

Month February Year 87 Page 1

CASH RECEIPTS JOURNAL

Description/ Explanation	Date	DEBITS Bank 1 Disbursement Account 11110	DEBITS Bank 2 Investment Account 11140	CREDITS Accounts Receivable 11320	CREDITS Miscellaneous Income 73000	CREDITS Miscellaneous Account #	CREDITS Miscellaneous Job #	CREDITS Miscellaneous Amount
John Smith	2/7	15,000:00		15,000:00				
ABC Partnership	2/8	320,150:00		320,150:00				
Jones Co. (joint ad)	2/10	45:00				61410		45:00
Alan Mark	2/10	100:00				11250		100:00
ABC Partnership	2/15		50,500:00	50,000:00		51600	4	500:00
Newman Ltd.	2/20		800:00		800:00			
XYZ Co.	2/25		100,000:00	100,000:00				
Totals		335,295:00	151,300:00	485,150:00	800:00			645:00

Figure 2.2

Accounts Payable

Accounts payable processing includes receiving invoices, matching them to purchase orders, coding to appropriate jobs and/or accounts, submission for approval, and entry into the system. Output involves the selection of which invoices are to be paid, followed by the generating of checks, which are then matched with invoices and purchase orders, submitted for signature and disbursed. The general ledger, accounts payable ledger, and job cost accounts are then updated and all data is filed.

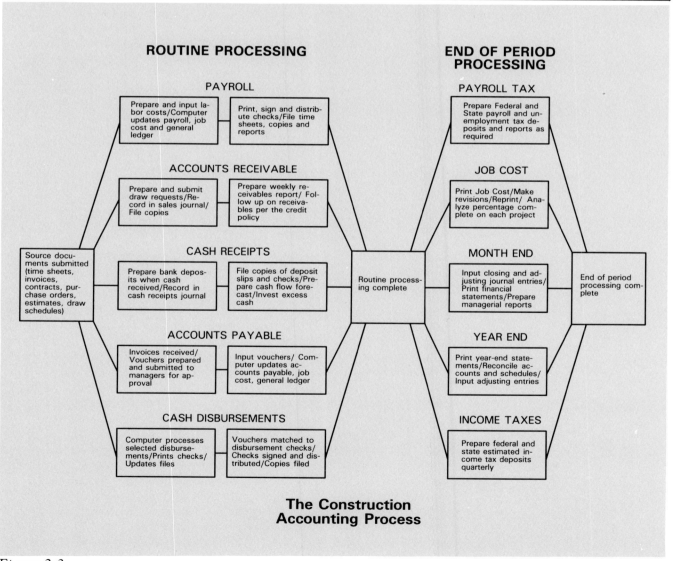

ROUTINE PROCESSING

END OF PERIOD PROCESSING

PAYROLL

Prepare and input labor costs/Computer updates payroll, job cost and general ledger

Print, sign and distribute checks/File time sheets, copies and reports

ACCOUNTS RECEIVABLE

Prepare and submit draw requests/Record in sales journal/File copies

Prepare weekly receivables report/ Follow up on receivables per the credit policy

CASH RECEIPTS

Prepare bank deposits when cash received/Record in cash receipts journal

File copies of deposit slips and checks/Prepare cash flow forecast/Invest excess cash

ACCOUNTS PAYABLE

Invoices received/ Vouchers prepared and submitted to managers for approval

Input vouchers/ Computer updates accounts payable, job cost, general ledger

CASH DISBURSEMENTS

Computer processes selected disbursements/Prints checks/ Updates files

Vouchers matched to disbursement checks/ Checks signed and distributed/Copies filed

Source documents submitted (time sheets, invoices, contracts, purchase orders, estimates, draw schedules)

Routine processing complete

PAYROLL TAX

Prepare Federal and State payroll and unemployment tax deposits and reports as required

JOB COST

Print Job Cost/Make revisions/Reprint/ Analyze percentage complete on each project

MONTH END

Input closing and adjusting journal entries/ Print financial statements/Prepare managerial reports

YEAR END

Print year-end statements/Reconcile accounts and schedules/ Input adjusting entries

INCOME TAXES

Prepare federal and state estimated income tax deposits quarterly

End of period processing complete

The Construction Accounting Process

Figure 2.3

Accounts Receivable

A major accounts receivable function is gathering all data to prepare a draw request or billing. This data consists of labor expenses, material invoices, subcontractor billings, equipment rental invoices, and miscellaneous invoices for items such as permit fees and assessments. The type of request for payment used depends on the contract for the project. If the contract is for a *fixed price*, draw requests should be submitted on a regular basis, based on the percentage of completion of the project. Billings for *time and materials* contracts should be generated on a regular interim basis, with the amount determined by expenses to date with applicable markups for operating expenses and profit.

Billings are recorded as *a credit to the sales account* and *debit to the accounts receivable account* through posting to the sales journal. As cash is received, deposits must be made and the accounts receivable ledger account updated. Collection of delinquent accounts in accordance with the firm's policy is another task within the accounts receivable segment.

Job Cost Processing

Job cost processing is affected by all other applications. For example, payroll processing updates the labor costs to each job; invoices recorded in accounts payable are coded as expenses to the appropriate job; and accounts receivable records revenue from each applicable job. Generally, job cost processing, independent of the other applications, is an *end of period* task. Job cost budgets should be adjusted each month in accordance with current projected estimates and reprinted to accurately state the percentage of completion on each project.

General Ledger

The general ledger consists of all the accounts in the Chart of Accounts. It is used to generate the financial statements. General ledger processing is primarily a culmination of all other segments.

Adjustments: Payroll, accounts payable and disbursements, accounts receivable and receipts, and job cost transactions all update the general ledger. The general ledger must be adjusted, however, at period end and cycle end. These adjustments are made to meet the requirements of an accrual basis of accounting. Accrual accounting attempts to more closely match a period's earned revenue with its incurred expenses, regardless of whether cash has been disbursed or received at that time.

Adjustments generally fall into one of two broad categories: *deferrals* and *accruals*. Deferral adjustments are generally necessary when there are prepaid expenses, when revenue is collected before it is earned, and for long term cost apportionment. Examples of deferral adjustments include prepaid insurance adjustments to recognize the period's insurance expense, revenue deferral for money advanced prior to the start of the project, and depreciation adjustment to recognize a portion of the expense for a fixed asset. Accrual adjustments are made for revenues earned but not yet collected, and for expenses incurred but not yet paid. Accounts receivable and payable are accruals, and involve period end adjustments, such as percentage of completion adjustments and accrued liabilities for payroll. At year end, there are additional adjustments, such as transfer of net profit or loss to the *retained earnings* account on the balance sheet. This procedure closes the income statement accounts so that there is no beginning balance for the new cycle.

The result of all of these tasks is a collection of *financial accounting reports*, including the Balance Sheet, Income Statement, and Statement of Cash Flows. These reports are covered in detail later in the chapter. In addition to these financial accounting reports, information gathered from the firm's financial records is also used in the preparation of managerial accounting reports. These reports summarize, forecast, or reformat accounting data such that management can better understand and use it for current and future planning. Some of the most common and informative managerial accounting reports are *operating budgets, job status reports, cash flow forecasts*, and *financial analysis reports*.

Accounting System Limitations

There are a number of limiting factors that affect the accuracy or usefulness of a contractor's accounting reports. One major limitation is the fact that financial reports relate only to *historical events and transactions*. Information needed by managers for *current control* is therefore not readily available. Managers must enlist their judgment and past experience in relating accounting reports to the present and projecting anticipated results into the future.

There is the potential for several system shortfalls. These can, however, be controlled by the manager who understands the system and the type of information that is needed. Some examples are outlined below.

1. *Inaccuracies in the Chart of Accounts.* If data is recorded in insufficient detail or classified improperly, the resulting financial statements will not report the firm's true performance. An analysis or comparison of such statements is misleading, causing management and external parties to arrive at incorrect conclusions. The Chart of Accounts must be designed specifically for general contractors, and must be capable of producing the types of reports included in this book. An excellent example of such a chart is shown in Appendix A.1.

2. *An inflexible or overly generalized job cost system.* Contractors may want to look to automated systems for the solution to this problem. Some of the more sophisticated systems are capable of accumulating and organizing data into practically any format and level of detail. An adequate job cost system is crucial due to the different types of projects encountered and the associated variations in cost structure.

3. *Clerical and procedural errors.* Most systems are subject to a multitude of human errors. Examples include:
   a. Small tools and supplies held in inventory. These items are often not charged properly to specific projects or allocated to all projects.
   b. An invoice may be paid twice.
   c. Labor, material, or subcontractors may not be posted to the proper job or cost code.
   d. Transpositions or account posting may be done incorrectly.

   Policies and procedures that can help avoid these types of problems are covered in Appendix A.2.

4. *Inaccurate collection of source documents from the field.* This process must be carried out quickly, but accurately. Therefore, clear procedures are required, with checks and balances. An example of the kinds of policies and procedures that are needed are included in Appendix A.3 and A.4, covering Purchasing and Field Documentation.

5. *Infrequent Statements.* Many smaller firms produce a statement only quarterly or annually. Using this method, periodic operating expenses paid only on occasion may not be accounted for on a statement. As a result, the records may fail to reflect the "true cost" as of the moment. It is important that most businesses produce current statements *every month* and record all accrued or prepaid expenses. In this way, an accurate picture of performance is available for management decisions. Monthly pro-rata charges for such items as depreciation, taxes, insurance, and interest must all be incorporated. Financial statements should be formatted with comparable data from the prior year for easy current period performance evaluation.

6. *Complexity and Inaccessibility.* Most accounting systems are so complex and handle so many items of data that it is very difficult for the contractor to determine where he stands quickly and easily. A system of *exception* or *variance* reporting must be developed whereby specific current shortfalls and problems in performance are readily apparent to management. The variance system should provide a clear trail for uncovering why things are not as originally planned and who is responsible. It is also beneficial to reduce the number of invoices received and checks issued, as outlined in Appendix A.3 on Purchasing. Purchase Orders and partial payments to subcontractors should be minimized, thereby limiting the data that must be managed.

Accounting Methods

There are four basic methods of accounting available to contractors. Two of these methods, *cash* and *accrual*, are choices available to most businesses. The other two, *percentage of completion* and *completed contract*, are the choices for the long term contract accounting generally found in construction. A long term contract is considered as one that requires more than 12 months to complete and/or overlaps the company's year end. The income recognition methods are different for long term contracts. This is because of the inherent difficulty in truly determining profitability on long term and complex projects that may be so affected by weather and other unknowns.

The choice of revenue recognition method affects both the firm's financial statement and its calculated taxable income. Unfortunately, a method that limits taxable income will have a negative impact on the firm's financial statements, including a smaller income, reduced net worth, and, possibly, less working capital. While the objective of tax avoidance might be achieved, the firm may subsequently be viewed less favorably by lending institutions and sureties.

Fortunately, this conflict can be avoided by utilizing separate methods of accounting for the two purposes. Many firms use the *cash* and *completed contract* methods for tax purposes while reporting for financial purposes on the *accrual* and *percentage-of-completion* methods. However, recent legislation has imposed limitations on the use of these methods. The tax implications of these methods are so great that they warrant their own detailed discussion in Chapter 8 on Tax Management. A brief discussion is provided here.

Cash

Using the cash basis method of accounting, revenue and expenses are recognized in the period in which cash is received or paid. Income is thereby calculated as the difference between cash collected and cash expended in any given period. Financial reporting using this method does not, in most cases, provide a true picture of financial position, since revenues and expenses are not matched. While this method can create tax advantages for the contractor, most third parties, such as lenders, will not accept financial statements generated on a cash basis.

Accrual

The goal of accrual basis accounting is to better match revenue earned in a period with the expenses incurred in that period, regardless of when cash has been received or paid. The accrual method adheres to the Generally Accepted Accounting Principles (G.A.A.P.). The G.A.A.P. is a set of standards developed by the Financial Accounting Standards Board (FASB), an independent nongovernmental body of accountants. This method provides a more useful picture of operating results than other methods. Most firms maintain their financial reporting records on an accrual basis. Even though this method may be less flexible for tax purposes, it may be required by the I.R.S., depending on the size of the firm.

The accrual method recognizes revenue when events occur that fix the *right to receive revenue* in an amount that can be determined accurately. For example, when the firm bills an owner, revenue will be recognized regardless of when it is collected. When materials are purchased or a subcontractor submits an invoice, the expense is realized even though it has not been paid. Income is thereby calculated as the difference between amounts billed to customers and liabilities for expenses incurred.

If amounts billed and uncollected are greater than expenses paid or incurred, the accrual method of accounting results in greater recognized income than that recognized by the cash method. However, if expenses paid or incurred exceeds amounts billed and uncollected, the cash method will result in greater recognized income than that of the accrual method.

Percentage of Completion

Percentage of completion is a method used to recognize revenue on long term construction contracts, or those expected to impact more than one fiscal year. This method is often used when reasonably accurate estimates can be made, contract rights are defined, and both the owner and contractor are expected to follow through with their respective obligations. This method recognizes gross profit, costs, and revenues throughout the project based on periodic measurements of progress. The gross margin is allocated to each accounting period based on the portion of the project estimated to be complete, which is the ratio of the current period's actual costs to the total estimated costs. For example, the total estimated cost of Project A is $1,000,000. The contract price is $1,200,000. During the first year of the project, the costs actually incurred totalled $750,000, or 75% of the total estimate. The revenue recognized for the first year is 75% of the contract price or $900,000, for a gross margin of $150,000. In the final period, the remaining actual gross margin is recognized.

The percentage of completion method is similar to the accrual method in that it recognizes income and expenses throughout the project. However, it differs in the case of overbillings. For example, if a contractor were to bill 50% of a $1,000,000 project when it is only 45% complete, he would be over-billing the customer by $50,000. The accrual method would recognize this $50,000 as income, while the percentage of completion method would recognize it as a liability. In the same manner, an underbilling would be recognized as an asset, using the percentage of completion method.

Completed Contract

Completed contract is a method of recognizing a project's revenue and expenses, only at completion. This method is commonly used for projects with unreliable estimates or other unknowns that preclude an accurate projection of profitability. While no profit is recognized until a project is complete, any known or expected losses must be reported immediately.

This method is not usually appropriate for financial reporting accuracy, for it fails to match current revenues with expenses. A contractor using the completed contract method might operate unprofitably on work-in-process, yet the true condition would not be evident from a review of his current financial statements. This method has primarily been used for tax purposes, and it is no longer available for larger contractors. See Chapter 8 for more detail.

Financial Statements

As discussed, the end results of the contractor's accounting function are statements that reflect the strength and performance of the firm. The most common statements are the *Balance Sheet*, *Income Statement*, and *Statement of Cash Flows*. Each will be briefly explained and illustrated with samples generated by a hypothetical contracting firm called *Eastway Contractors*. It may be helpful to read the 1988 Business Plan for Eastway Contractors, found in Appendix B.1, prior to studying these sample statements.

In addition to these financial statements, there are a number of notes that CPA's commonly include with financial statements. These notes explain conditions that are currently or may in the future impact the stability of the firm. Examples of such notes are:

- An explanation of all terms related to short and long term debt.
- An explanation of any contingent liabilities that may affect the firm.
- A discussion of the firm's obligations in any major employee benefit programs, such as pension or profit sharing plans.
- An explanation of the firm's overall tax exposure.
- A discussion of lease agreements binding the firm.
- An explanation of any joint venture agreements involving the firm.
- A statement of the effect of the accounting methods used by the firm on the accuracy of the statements.

Balance Sheet

The balance sheet is an expanded version of the following accounting equation:

$$\text{Assets} = \text{Liabilities} + \text{Equity}$$

The balance sheet is considered to be a statement of the firm's financial position as of the date it was prepared. Dollar sums representing all of the firm's assets, liabilities, and owner's equity (or net worth) are summarized, as shown in Figure 2.4. Equity represents the owner's investment in the firm versus that contributed by creditors, and can be computed by the following equation.

$$Equity = Assets - Liabilities$$

This broad-based economic equation is subdivided on the balance sheet into classifications and titles (as shown in the Chart of Accounts, Appendix A.1) that allow for an analysis and comparison of current financial statements with past statements and industry averages. The most common classification and titles are as follows:

Assets:

Current Assets
- Cash and Equivalents
- Receivables
- Inventories
- Prepaid Expenses

Capital (Non-Current) Assets
- Construction Equipment
- Trucks and Autos
- Office Equipment
- Real Estate

Deferred Items
- Taxes
- Other Charges

Liabilities:

Current Liabilities
- Payables
- Accrued Liabilities
- Payroll Withholdings
- Income Taxes Payable

Long Term Liabilities
- Notes Payable
- Deferred Taxes

Stockholders' Equity:
- Capital Stock
- Paid-In-Capital
- Retained Earnings

The makeup of the line items in a balance sheet such as that shown in Figure 2.4 is of great importance to the contractor. If specific assets or liabilities are not properly classified, the statement may be misleading and difficult to relate to industry standards. The importance of formatting and comparability is detailed in Chapter 3 on Financial Analysis. Listed below are the items included in each line item of a typical contractor's balance sheet. (In accordance with percentage-of-complete accounting, which is recommended for accurate financial reporting, the balance sheet includes an asset for underbillings and liability for overbilling.)

Current Assets:

Cash and Equivalents: All cash, marketable securities, and other near-cash items.

EASTWAY CONTRACTORS, INC.
BALANCE SHEET
DECEMBER 31, 1987 AND 1986

ASSETS

	1987	1986
Current Assets		
Cash	$295,000	$160,000
Accounts Receivable – Trade	400,000	320,000
Accounts Receivable – Retention	80,000	70,000
Inventory	10,000	15,000
Costs and estimated earnings in excess of billings	50,000	45,000
All Other Current Assets	25,000	10,000
Total Current Assets	860,000	620,000
Plant, Property and Equipment – at cost		
Construction Equipment	40,000	75,000
Trucks and Autos	30,000	30,000
Office Equipment	30,000	20,000
	100,000	125,000
Less: Accumulated depreciation and amortization	15,000	15,000
Plant, Property and Equipment, net	85,000	110,000
Other Assets		
Joint Ventures	0	15,000
Other Non-Current Assets	40,000	35,000
Total Other Assets	40,000	50,000
TOTAL ASSETS	$985,000	$780,000

Figure 2.4

LIABILITIES AND STOCKHOLDER'S EQUITY

	1987	1986
Current Liabilities		
Notes Payable	$30,000	$40,000
Accounts Payable – Trade	265,000	275,000
Accounts Payable – Retention	25,000	25,000
Billings in excess of costs and estimated earnings	60,000	50,000
Income Taxes Payable	25,000	10,000
Current Maturity – Long Term Debt	15,000	20,000
All Other Current Liabilities	25,000	23,000
Total Current Liabilities	445,000	443,000
Long Term Debt	30,000	40,000
Deferred Income Tax	25,000	5,000
Other Non-Current Liabilities	15,000	12,000
	70,000	57,000
Total Liabilities	515,000	500,000
Stockholders' Equity		
Common Stock	10,000	10,000
Retained Earnings		
Balance – beginning of period	280,000	200,000
Net Profit/(Loss)	180,000	70,000
Balance – end of period	460,000	270,000
Total Stockholders' Equity	470,000	280,000
TOTAL LIABILITIES AND STOCKHOLDERS' EQUITY	$985,000	$780,000

Figure 2.4 (*continued*)

Accounts Receivable Trade: Amounts billed and due on current contracts.

Accounts Receivable Retention: Amounts previously billed and withheld by customers on current contracts.

Inventory: Includes the following:
- Deferred costs for marketing materials, office supplies, and miscellaneous warranty supplies not normally allocated to a specific project.
- Materials purchased in bulk for numerous projects and not yet allocated to individual projects.
- Under the completed contracts method, inventory includes all direct costs incurred for projects not yet complete.

Costs and Estimated Earnings in Excess of Billings: The difference between the total costs and recognized estimated earnings to date and the total actual billings to date.

All Other Current: Any other current assets. Does not include prepaid items.

Total Current: Total of all assets listed above.

Noncurrent Assets:

Plant, Property, and Equipment: All property, plant, leasehold improvements, and equipment (fixed assets), net of accumulated depreciation or depletion.

Joint Ventures and Investments: The total of investments and equity in joint ventures.

All Other Noncurrent: Prepaid items, cash value of life insurance, and other noncurrent assets. This category also includes intangible assets if applicable, including goodwill, trademarks, franchises, and all net of accumulated amortization.

Total Assets: Total of all current and capital assets listed above.

Current Liabilities:

Notes Payable—Short Term: All short term note obligations, including bank lines of credit. Does not include trade notes payable.

Accounts Payable Trade: Financial obligations to suppliers, subcontractors, professional associates, and other creditors.

Accounts Payable—Retention: Amounts held back as retention in payments to subcontractors on current contracts.

Billing in Excess of Costs and Estimated Earnings: The difference between the total billings to date and the total of costs and recognized earnings to date.

Income Taxes Payable: Income tax obligations including the current portion of deferred taxes.

Current Maturities—Long Term Debt: The portion of long term obligations that are due within the current fiscal year.

All Other Current: Any other current liabilities, including bank overdrafts and accrued expenses.

Total Current: Total of all liabilities listed above.

Long Term Liabilities:

Long Term Debt: Bank debt, mortgages, deferred portions of long term debt, and capital lease obligations.

Deferred Taxes: Total of all deferred taxes.

All Other Noncurrent: Any other noncurrent liabilities, including subordinated debt, and liability reserves.

Total Liabilities: Total of all current and long term liabilities.

Net Worth (stockholders' equity): Difference between total assets and total liabilities.

Total Liabilities and Net Worth: Total of all items listed above.

Income Statement

This report consists of the revenue earned and expenses incurred for the accounting period. It reflects the activity of the firm for a specific time period, generally one month, year-to-date, or one year. While many small firms produce an income statement annually, it is highly advised that such a statement be prepared monthly. Such interim statements are of great value to management in assessing the firm's current performance compared with prior periods.

As the net assets of the firm change relative to its liabilities, a profit or loss is realized. This profit or loss is transferred each period to the *retained earnings* section of the balance sheet. An Income Statement for Eastway Contractors is shown in Figure 2.5, and a detailed description of each line item is provided below.

Contract Revenues: Total annual revenues recognized under the completed contract method (generally preferred for tax reporting) or the percent of completion method (generally preferred for financial reporting) for construction and design services.

Direct Costs: Costs incurred in the field for all labor, material, and subcontractor services expended in the actual construction of projects. All costs for temporary utilities such as field offices, power, and phone service would ideally be job-costed as a direct cost of each project. Small tools, supplies, and rental equipment used on a project are also considered direct costs, if they are easily identifiable with a specific project.

Direct labor cost includes the hourly wages and/or salaries of field employees. Other costs associated with field employees, such as benefits, Workers Compensation, and Unemployment Insurance may be considered indirect costs and allocated to all projects built during the year. Larger contractors building projects of long duration may choose to direct-cost these items if they can be related to a single project.

Direct design costs, although not typically job-costed to a particular project, are considered a direct cost against design revenues.

Indirect Costs: Costs incurred as the result of field operations, but which are not direct costs (labor, material, or subcontractor expenses). Indirect costs include the following:

- Benefits, Workers Compensation, Unemployment Insurance, and bonuses for field employees, including salaried supervisors.
- Tools, trucks, and equipment (including depreciation) used on numerous projects, as well as labor required for routine maintenance of such equipment.
- Nonrecoverable engineering and architectural expenses.
- Warranty expense.
- Builders Risk Insurance.

EASTWAY CONTRACTORS, INC.
INCOME STATEMENT
FOR THE YEARS ENDED DECEMBER 31,

	1987		1986	
	AMOUNT	% OF INCOME	AMOUNT	% OF INCOME
REVENUE				
Construction Contracts	$5,950,000	99.17%	$3,985,000	99.63%
Design Contracts	50,000	0.83%	15,000	0.38%
TOTAL REVENUE	6,000,000	100.00%	4,000,000	100.00%
COST OF REVENUE				
DIRECT COST	5,140,000	85.67%	3,455,000	86.38%
INDIRECT COSTS				
Field Employees	20,000	0.33%	25,000	0.63%
Field Equipment	16,000	0.27%	18,000	0.45%
Field Insurance	25,000	0.42%	20,000	0.50%
Field Vehicles	25,000	0.42%	30,000	0.75%
Warranty	20,000	0.33%	27,000	0.68%
Other Indirect Expenses	24,000	0.40%	15,000	0.38%
TOTAL INDIRECT COSTS	130,000	2.17%	135,000	3.38%
TOTAL COST OF REVENUE	5,270,000	87.83%	3,590,000	89.75%
GROSS PROFIT	730,000	12.17%	410,000	10.25%
OPERATING EXPENSES				
Business Insurance	7,000	0.12%	5,000	0.13%
Occupancy	45,000	0.75%	35,000	0.88%
Office Employees	140,000	2.33%	100,000	2.50%
Office Equipment	16,000	0.27%	10,000	0.25%
Office Supplies	15,000	0.25%	11,000	0.28%
Office Vehicles	15,000	0.25%	8,000	0.20%
Officer Compensation	150,000	2.50%	90,000	2.25%
Professional Fees	12,000	0.20%	6,000	0.15%
Sales & Marketing	35,000	0.58%	15,000	0.38%
Travel & Entertainment	10,000	0.17%	7,000	0.18%
Other Gen. & Admin. Expenses	35,000	0.58%	21,000	0.53%
TOTAL OPERATING EXPENSES	480,000	8.00%	308,000	7.70%
OPERATING PROFIT	250,000	4.17%	102,000	2.55%
ALL OTHER INCOME & EXPENSE (NET)	10,000	0.17%	2,000	0.05%
PROFIT BEFORE TAX	260,000	4.33%	104,000	2.60%
INCOME TAX	80,000	1.33%	34,000	0.85%
NET PROFIT	$180,000	3.00%	$70,000	1.75%

Figure 2.5

EASTWAY CONTRACTORS, INC.
SCHEDULE OF INDIRECT COSTS
1987 AND 1986 INCOME STATEMENT

	1987	1986
FIELD EMPLOYEE EXPENSES		
Holidays	$1,000	$2,700
Insurance – Health	3,000	4,500
Retirement	2,000	3,000
Salaries-Bonus	11,000	9,000
Salaries-Vacation, Sick, Other	3,000	5,800
TOTAL FIELD EMPLOYEE EXPENSES	20,000	25,000
FIELD EQUIPMENT EXPENSES		
Depreciation	7,500	8,500
Rental	5,000	6,000
Repairs	3,500	3,500
TOTAL FIELD EQUIPMENT EXPENSES	16,000	18,000
FIELD INSURANCE EXPENSES		
Builders' Risk	12,500	10,000
General Liability	12,500	10,000
TOTAL FIELD INSURANCE EXPENSES	25,000	20,000
FIELD VEHICLE EXPENSES		
Depreciation	7,500	8,000
Fuel & Oil	7,000	7,500
Insurance	6,000	7,000
Repairs & Maintenance	4,500	7,500
TOTAL FIELD VEHICLE EXPENSES	25,000	30,000
WARRANTY EXPENSE	20,000	27,000
OTHER INDIRECT EXPENSES		
Miscellaneous	14,500	8,500
Small Tools & Supplies	7,500	5,500
Travel	2,000	1,000
TOTAL OTHER INDIRECT EXPENSES	24,000	15,000
TOTAL INDIRECT COSTS	$130,000	$135,000

Figure 2.5 (continued)

EASTWAY CONTRACTORS, INC.
SCHEDULE OF OPERATING EXPENSES
1987 AND 1986 INCOME STATEMENT

	1987	1986
BUSINESS INSURANCE		
Building	3,000	3,000
Liability & Office Equipment	4,000	2,000
Other	0	0
TOTAL BUSINESS INSURANCE	7,000	5,000
OCCUPANCY EXPENSES		
Rent	5,000	3,500
Repairs & Maintenance	30,000	24,000
Taxes	1,000	1,000
Telephone	5,500	4,000
Utilities	3,500	2,500
TOTAL OCCUPANCY EXPENSES	45,000	35,000
OFFICE EMPLOYEE EXPENSES		
Continuing Education	3,000	1,700
Insurance – Health	3,500	2,000
Insurance – Workers Comp.	1,000	400
Payroll Taxes	15,000	9,500
Retirement	1,500	1,100
Salaries–Bonus	7,000	2,000
Salaries–Staff	107,500	82,500
Social Activities	1,500	800
TOTAL OFFICE EMPLOYEE EXPENSES	140,000	100,000
OFFICE EQUIPMENT EXPENSES		
Depreciation	8,000	7,000
Rental	4,500	2,000
Repairs	3,500	1,000
TOTAL OFFICE EQUIPMENT EXPENSES	16,000	10,000
OFFICE SUPPLIES	15,000	11,000
OFFICE VEHICLE EXPENSES		
Depreciation	6,500	3,500
Fuel & Oil	5,000	3,100
Insurance	2,000	1,200
Repairs & Maintenance	1,500	200
TOTAL OFFICE VEHICLE EXPENSES	15,000	8,000
OFFICERS COMPENSATION		
Insurance-Disability	1,000	1,000
Insurance-Life	1,500	1,500
Salaries–Bonus	0	0
Salaries–Officers	147,500	87,500
Retirement	0	0
TOTAL OFFICERS COMPENSATION	150,000	90,000

Figure 2.5 (continued)

EASTWAY CONTRACTORS, INC.
SCHEDULE OF OPERATING EXPENSES
1987 AND 1986 INCOME STATEMENT

	1987	1986
PROFESSIONAL FEES		
Accounting	7,000	2,000
Legal	4,000	3,000
Other	1,000	1,000
TOTAL PROFESSIONAL FEES	12,000	6,000
SALES & MARKETING EXPENSES		
Advertising	22,000	10,000
Commissions	0	2,000
Literature	4,000	2,000
Promotions	9,000	1,000
SALES & MARKETING EXPENSES	35,000	15,000
TRAVEL & ENTERTAINMENT EXPENSES		
Meals	3,000	2,500
Travel	7,000	4,500
TOTAL TRAVEL & ENTERTAINMENT EXPENSES	10,000	7,000
OTHER GEN. & ADMIN. EXPENSES		
Bad Debt	7,000	5,600
Branch Office Expense	4,000	3,500
Classified Advertising	6,500	3,000
Contributions	7,500	3,000
Dues & Subscriptions	500	900
Fines & Penalties	500	500
Licensing	1,500	1,500
Miscellaneous	3,500	2,000
Temporary Secretarial	4,000	1,000
TOTAL GEN. & ADMIN. EXPENSES	35,000	21,000
TOTAL OPERATING EXPENSES	480,000	308,000

DETAILS TO ALL OTHER INCOME & EXPENSE (NET)

	1987	1986
OTHER INCOME		
Interest Income	6,000	2,000
Miscellaneous Income	12,000	5,000
OTHER EXPENSE		
Interest Expense	(8,000)	(5,000)
TOTAL ALL OTHER INCOME & EXPENSE (NET)	10,000	2,000

Figure 2.5 (continued)

Gross Profit: The difference between contract revenues and the sum of direct and indirect costs (cost of sales)

Operating Expenses: All marketing and general and administrative expenses, including:

- Salaries and benefits of administrative employees
- Office rent, utilities, maintenance
- Car expenses, depreciation
- Business insurance
- Accounting, legal, and other professional services
- Office supplies, equipment, repairs
- Depreciation of fixed non-field-related assets
- Dues, subscriptions, contributions
- Travel and entertainment
- Marketing expenses

Operating Profit: Gross profit minus operating expenses.

All Other Income and Expenses (net): Includes other income and expenses (net), such as interest expense, miscellaneous expenses not included in general & administrative expenses, netted against such recoveries as interest income, dividends received, and miscellaneous income.

Profit Before Taxes: Operating profit plus all other income and expenses (net).

Income Tax: Taxes incurred at the federal, state, or local level.

Net Profit: Profit remaining after payment of taxes.

Statement of Cash Flow

This report summarizes the net change in the firm's cash position during a given period. In essence, the net cash (gained or lost) through operations, investing, and financing are all combined on this single report. A sample statement of cash flow for Eastway Contractors is provided in Figure 2.6.

The Job Cost System

Job cost systems are becoming increasingly sophisticated due in large part to rapidly increasing construction costs and competition. These external influences make up-to-date information for management decision making a critical ingredient for success. Without an adequate job cost system, profitability on a project is virtually impossible to determine. Maintaining accurate historical cost and material quantity data is also very important for use in estimating future projects.

A job cost system can be thought of as a series of *source documents and procedures* used to record, process, and format job cost information for project management decisions. Many different types of systems can be used, from a simple manual one-write system to a highly sophisticated multi-modular computerized system. Regardless of the type of system most appropriate for a firm, the objectives of a job cost system are the same. It should identify all direct costs traceable to the project in a format consistent with the firm's estimating structure, thereby allowing a comparison of estimated versus actual costs. Any variances should be readily apparent such that management can respond as required. The most important benefits of a job cost system include:

1. Job Cost Control.
2. Historical Information for Estimating Future Jobs.
3. Analysis of Profitability by Job.
4. Analysis of Profitability by Profit Centers or Type of work.

EASTWAY CONTRACTORS, INC.
STATEMENT OF CASH FLOWS
FOR THE YEAR ENDED DECEMBER 31, 1987

Cash flows from operating activities:
  Cash received from customers $5,910,000
  Interest received   6,000

    Cash provided by operating activities   5,916,000

  Cash paid to suppliers and employees   5,700,000
  Interest and taxes paid   98,000

    Cash disbursed for operating activities   5,798,000

Net cash flow from operating activities   118,000

Cash flows from investing activities:
  Proceeds from sale of plant assets   18,000
  Proceeds from sale of joint venture   24,000

Net cash provided by investing activities   42,000

Cash flows from financing activities:
  Proceeds from short term borrowing   10,000
  Payments to settle short term debts   (25,000)
  Payments to settle long term debts   (10,000)

Net cash used by financing activities   (25,000)

Net increase in cash $135,000

Beginning Cash Balance   160,000

Ending Cash Balance $295,000

Figure 2.6

Clearly, the amount of detailed information available for the above depends on the sophistication of the system. An advanced integrated job cost system is illustrated in Figure 2.7. This diagram shows the multiple interactions between the job cost system and the other financial functions of the firm.

Establishing a record for each project or job, whether it is a card in a manual system or a computer record, is the first step in creating a job cost system. This record must span the entire life of the project, rather than just one accounting cycle. At the onset of planning for a project, an estimate summary of costs for each cost category must be prepared,

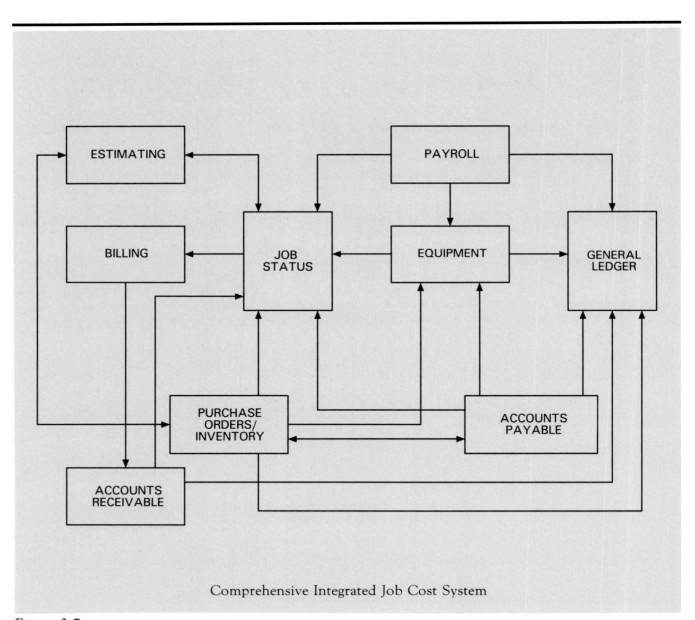

Comprehensive Integrated Job Cost System

Figure 2.7

dividing the estimated costs, hours, or production units into classifications such as *material*, *labor*, *subcontractor*, *equipment*, and *miscellaneous*. A sample form for an Estimate Summary that conforms to the job cost accounts can be found in Appendix A.5.

All invoices that can be identified with a particular job must be posted to the job record. These costs, in addition to the estimates, provide the data to generate job cost reports. Figure 2.8 is a job cost report generated using *COINS, Construction Industry Software*, a product of Shaker Computers. This example of a job cost format reports the original estimate (OE), the present adjusted estimate (PE) and the actual costs (ACT) to date for each type of direct cost for each cost code, along with the variance if actual costs exceed the present estimate. The present estimate (PE) is a sum that may constantly change in response to a number of project variables, including more favorable purchasing, change orders, or improved productivity. These variables must be routinely updated such that the percentage-of-complete computation is accurate.

This approach of constant updating is sometimes referred to as *Indicated Outcome Contract Accounting (IOCA)*. Under IOCA, a contractor looks continuously at the projected gross profit on each project. Any events that are likely to affect the indicated outcome, such as variables affecting the present estimate or contract amount, are taken into account. While IOCA is not very complex or costly, it does require careful planning in setting up the system at the outset. The rewards are great, however, in terms of accurate and valuable cost reports.

At the end of the report, *contract amounts* and *amount billed* are reported, as well as information concerning the amount of anticipated gross margin and to-date information. These reports enable the manager to review the actual costs versus the anticipated costs and to adjust the estimates as necessary.

Job Status Reports

A Job Status Report, an end product of a job cost system, is a most useful tool for management analysis and decision making. Some systems title this report *Percentage of Completion Schedule*, but the contents and objectives are the same. Figure 2.9 illustrates one format for this report. While the presentation format may differ from system to system, the basic information provided should be the same.

Contract Amount: the original contract plus all change orders to date.

Present Estimate: the original cost estimate plus all adjustments.

Estimated Profit: the difference between the contract amount and the estimated costs.

Profit Percentage: the estimated profit divided by the estimated costs.

Costs to Date: the actual costs entered into the system as of the current period.

Percent complete: determined by dividing the actual costs to date by the estimated total costs.

Earnings to date: the percent complete times the contract amount.

Profit to date: the earnings to date less the actual costs to date. The profit is described in terms of the year in which it was earned (in prior years or in the current year).

EASTWAY CONTRACTORS, INC.

JOB: 8   Sample Project

COST CODE	DESCRIPTION		LABOR	MATERIAL	SUBCONTRCT	EQUIPMENT	MISC.		TOTAL COSTS	CONTRACT AMOUNT// AMOUNT BILLED	CONT-ACT/ PE - ACT/ BILL-ACT/ CONT - PE
51220	Temp. Serv.	OE	300	500	7200	500	0	0	8500		
		PE	700	1200	5700	400	0	0	8000		
		ACT	640	1135	5000	300	0	0	7075		
51230	Cleaning	OE	2000	0	6000	0	0	0	8000		
		PE	2000	0	6000	0	0	0	8000		
		ACT	500	0	3000	0	0	0	3500		
51240	Misc.Gen.	OE	0	500	1500	0	500	0	2500		
		PE	0	500	1500	0	500	0	2500		
		ACT	0	0	350	0	500	0	850		
51270	Earthwork	OE	2000	0	250000	0	0	0	252000		
		PE	1800	0	260000	0	0	0	261800		
		ACT	1600	0	230000	0	0	0	231600		
51300	Utilities	OE	0	3000	10000	0	0	0	13000		
		PE	0	3000	10000	0	0	0	13000		
		ACT	0	1250	7500	0	0	0	8750		
51320	Curb/Gutter	OE	1500	0	20000	0	0	0	21500		
		PE	1500	0	20000	0	0	0	21500		
		ACT	1000	0	0	0	0	0	1000		
51330	Site Light.	OE	0	2500	3000	0	0	0	5500		
		PE	0	2500	3000	0	0	0	5500		
		ACT	0	2650	3000	0	0	0	5650	OVER ESTIMATE	150
51340	Signage	OE	0	0	18000	0	1000	0	19000		
		PE	0	0	16500	0	1000	0	17500		
		ACT	0	0	0	0	0	0	0		
51350	Seed/Landsc	OE	0	0	25000	0	0	0	25000		
		PE	0	0	23000	0	0	0	23000		
		ACT	0	0	0	0	0	0	0		
51360	Misc. Site	OE	1000	7500	45000	750	550	0	54800		
		PE	1000	7500	45000	750	550	0	54800		
		ACT	150	260	38500	750	400	0	40060		
51380	Footings	OE	0	0	30000	0	0	0	30000		
		PE	0	0	28900	0	0	0	28900		
		ACT	0	0	28850	0	0	0	28850		

Figure 2.8

EASTWAY CONTRACTORS, INC.

JOB: 8   Sample Project

COST CODE	DESCRIPTION		LABOR	MATERIAL	SUBCONTRCT	EQUIPMENT	MISC.		TOTAL COSTS	CONTRACT AMOUNT // AMOUNT BILLED	CONT-ACT/ PE - ACT/ BILL-ACT/ CONT - PE
51390	Slabs	OE	1500	2500	65000	0	0	0	69000		
		PE	1500	2500	65000	0	0	0	69000		
		ACT	1700	2100	65500	0	0	0	69300	OVER ESTIMATE	300
51440	Veneer	OE	0	0	225000	0	0	0	225000		
		PE	0	0	225000	0	0	0	225000		
		ACT	0	0	215000	0	0	0	215000		
51460	Structural	OE	0	0	50000	0	0	0	50000		
		PE	0	0	50000	0	0	0	50000		
		ACT	0	0	40000	0	0	0	40000		
51490	Fasteners	OE	0	0	2500	0	0	0	2500		
		PE	0	0	2500	0	0	0	2500		
		ACT	0	0	2000	0	0	0	2000		
51510	Rough	OE	10000	110000	60000	1000	1200	0	182200		
		PE	9000	108000	60500	1000	1200	0	179700		
		ACT	8900	107500	50000	700	500	0	167600		
51520	Trim/Stairs	OE	0	0	10000	0	0	0	10000		
		PE	0	0	10000	0	0	0	10000		
		ACT	0	0	0	0	0	0	0		
51550	Termite/Wtr	OE	0	0	7500	0	0	0	7500		
		PE	0	0	7500	0	0	0	7500		
		ACT	0	0	7200	0	0	0	7200		
51560	Insulation	OE	0	0	45000	0	0	0	45000		
		PE	0	0	41500	0	0	0	41500		
		ACT	0	0	40000	0	0	0	40000		
51570	Roofing	OE	0	0	100000	0	1800	0	101800		
		PE	0	0	100000	0	1800	0	101800		
		ACT	0	0	110000	0	1760	0	111760	OVER ESTIMATE	9960
51580	Siding/Ext.	OE	500	2000	40000	800	0	0	43300		
		PE	1000	2750	35000	800	0	0	39550		
		ACT	850	2500	12000	860	0	0	16210		
51600	Metal/Spec.	OE	2000	10000	0	0	0	0	12000		
		PE	2000	10000	0	0	0	0	12000		
		ACT	1000	5000	0	0	0	0	6000		

Figure 2.8 (continued)

EASTWAY CONTRACTORS, INC.

JOB: 8   Sample Project

COST CODE	DESCRIPTION		LABOR	MATERIAL	SUBCONTRCT	EQUIPMENT	MISC.		TOTAL COSTS	CONTRACT AMOUNT// AMOUNT BILLED	CONT-ACT/ PE - ACT/ BILL-ACT/ CONT - PE
51610	Windows	OE	8000	30000	0	0	0	0	38000		
		PE	7250	27000	0	0	0	0	34250		
		ACT	6500	24500	0	0	0	0	31000		
51620	Glazing Sys	OE	0	0	60000	0	0	0	60000		
		PE	0	0	52500	0	0	0	52500		
		ACT	0	0	30000	0	0	0	30000		
51640	Drywall	OE	2500	5000	50000	0	500	0	58000		
		PE	2500	5000	50000	0	500	0	58000		
		ACT	2550	5200	40000	0	0	0	47750		
51660	Acoustical	OE	0	0	45000	0	0	0	45000		
		PE	0	0	40000	0	0	0	40000		
		ACT	0	0	8000	0	0	0	8000		
51670	Painting	OE	0	0	15000	0	0	0	15000		
		PE	0	0	12000	0	0	0	12000		
		ACT	0	0	1200	0	0	0	1200		
51680	Specialties	OE	3000	5000	16000	1000	0	0	25000		
		PE	3000	5000	16000	1000	0	0	25000		
		ACT	1200	1500	8700	0	0	0	11400		
51760	Plumbing	OE	0	0	45000	0	0	0	45000		
		PE	0	0	40000	0	0	0	40000		
		ACT	0	0	27500	0	0	0	27500		
51770	HVAC	OE	0	0	120000	0	0	0	120000		
		PE	0	0	102700	0	0	0	102700		
		ACT	0	0	85000	0	0	0	85000		
51790	Serv/Wiring	OE	0	0	125000	0	2800	0	127800		
		PE	0	0	125000	0	3000	0	128000		
		ACT	0	0	85000	0	3000	0	88000		
JOB TOTALS		OE	34300	178500	1496700	4050	8350	0	1721900	1900000	667745
		PE	33250	174950	1454800	3950	8550	0	1675500	2000000	343245
		ACT	26590	153595	1143300	2610	6160	0	1332255	1500000	167745
											324500

Figure 2.8 (continued)

EASTWAY CONTRACTORS, INC.

COINS JOB COST SYSTEM
PERCENTAGE OF COMPLETION SCHEDULE
THRU PERIOD ENDING 09/30/89
COMPLETE & INCOMPLETE JOBS
ALL GROUPS
ALL TYPES

JOB NAME	JOB #	CONTRACT AMOUNT A	PRESENT ESTIMATE B	ESTIMATED PROFIT C	% PROFIT D (C/B)	COSTS TO DATE B	% COMPLETE F (B/B)	EARNINGS TO DATE G (F*A)	PROFIT TO DATE H (G-B)	PROFIT PRIOR YEARS I	PROFIT THIS YEAR J (H-I)	BILLINGS TO DATE K	UNDER BILLED L (G-K)	OVER BILLED M (K-G)	CONTRACT BACKLOG N (A-G)	COST TO COMPLETE O (B-B)	FUTURE PROFIT P (C-H)
OFFICE CONDOS	1	1590000	1333800	256200	19.21	1000000	74.97	1192023	192023	96091	95932	1110000	82023	0	397977	333800	64177
AUTO MALL	11	1200000	1030000	170000	16.50	0	0.00	0	0	0	0	0	0	0	1200000	1030000	170000
RETAIL REMODEL	13	600000	498000	102000	20.48	0	0.00	0	0	0	0	0	0	0	600000	498000	102000
SELF STORAGE	14	700000	590000	110000	18.64	0	0.00	0	0	0	0	0	0	0	700000	590000	110000
TOWNHOUSES	2	1540000	1300000	240000	18.46	1075000	82.69	1273426	198426	124568	73858	1285000	0	11574	266574	225000	41574
APARTMENTS	3	1015000	880000	135000	15.34	880000	100.00	1015000	135000	46014	88987	925000	90000	0	0	0	0
RETAIL CENTER	4	2010000	1675000	335000	20.00	550000	32.84	660084	110084	50093	59991	675000	0	14916	1349916	1125000	224916
OFFICE WAREHOUSE	5	1575000	1336000	239000	17.89	950000	71.11	1119983	169983	62650	107333	1110000	9983	0	455018	386000	69018
TOWNHOUSES II	6	1000000	844200	155800	18.46	350000	41.46	414600	64600	0	64600	420000	0	5400	585400	494200	91200
REPORT TOTALS		11230000	9487000	1743000	18.37	4805000	50.65	5675116	870116	379416	490700	5525000	182006	31890	5554885	4682000	872885

Figure 2.9

Billings to Date: the actual draw requests entered into the system. Underbillings (unbilled accounts receivable) and overbillings (billed accounts receivable) in excess of earnings are determined by subtracting the actual amount billed from the amount earned.

Contract Backlog: the amount remaining to be earned on the contract.

Costs to Complete: total estimated costs less costs to date.

Future Profit: the difference between the *estimated total profit* and the *profit to date*.

Benefits: The Job Status, or Percentage of Completion report serves a dual purpose for management. First, it provides a "bird's eye view" of the profitability of each project in an easy to understand format. Second, in order to use the percentage of completion method of accounting for long term projects, adjustments must be made such that financial statements will reflect the true earned revenue to date. This report provides the amount of the adjustment in the under- and over-billed columns. The underbilled amount is an asset and an adjustment which increases revenue. The overbilled amount is a liability which reduces recognized revenue. Both of these adjustments are reflected in the Balance Sheet shown in Figure 2.4.

Potential Problems: Any discussion of job cost would not be complete without acknowledging the problems that may arise in attempting to establish and utilize a viable system. Complications may result from the multiple sources of data: information generated by field personnel, the owner, subcontractors, vendors, and the office. Establishing effective channels of communication as well as sound policies and procedures is an important step toward minimizing problems with data collection.

The multiple *types* of data add to the challenges of construction cost reporting. With so many items, including estimated costs, backcharges, change orders, delivery tickets, time reporting, and quantities, it is easy to become bogged down. However, establishing a job cost system need not be an overwhelming project. Furthermore, doing without such a system may mean financial ruin for the firm, particularly in these times of rapidly fluctuating prices and prolific competition.

Summary

Construction accounting is a complex yet crucial factor in the success of a contracting operation. Effective managers must understand the basic mechanics of their accounting system and the type of output it generates. Furthermore, the output must be formatted such that it can be easily and quickly reviewed, with variances highlighted for quick recognition. Not only management, but many parties outside the firm, require current and accurate financial data on the firm's strength and performance.

In addition to the standard financial accounting reports, such as the Income Statement, Balance Sheet, and Statement of Cash Flows, financial accounting data is also used in a number of managerial accounting reports, such as budgets, cash flow projections, and financial ratios.

Construction accounting brings challenges and considerations different from those of many other businesses. First, there are four accounting methods available to contractors: *cash*, *accrual*, *completed contract*, and *percentage of complete*. Each of these methods may have very different tax consequences. Therefore, the guidance of a CPA or tax attorney is strongly recommended. Maintaining accurate financial reports is another special challenge to mangers of construction firms, particularly when one considers the potential for over- or underbillings on contract sums. Finally, the job cost system so important in contracting requires a special set of procedures and control. In all cases, there are certain accounting system limitations peculiar to construction businesses, all of which must be addressed as effectively as possible.

Chapter Three

FINANCIAL ANALYSIS

Chapter Three

FINANCIAL ANALYSIS

The section entitled, "Why Contractors Fail," in Chapter 1 identified a number of the major causes of bankruptcy among construction firms. While many managers tend to look first at field production as the likely source of their problems, shortcomings in the firm's financial structure and control systems are more often the cause of failure. Even those firms with outstanding field control and contract prices at potentially profitable levels can still run into financial difficulties due to a wide range of other factors.

Effective techniques have been developed to assist managers in identifying and analyzing the true causes of their firm's financial shortfalls. The two most common approaches to evaluating a firm's financial strength involve the analysis of financial *ratios and trends*, both of which are explained in detail in this chapter. Each of these methods simplifies and summarizes key financial data in a form that allows managers to compare it with prior periods, industry standards, and goals established by the firm.

Benefits of Financial Analysis

The objective of financial analysis is to provide management with a clear understanding of both the positive and negative aspects of the firm's financial structure and performance. With this understanding, contractors are able to respond to negative indicators long before serious problems develop. Through consistent attention to financial analysis and the appropriate corrective actions, a building firm can ultimately become a finely tuned "Strike Force", capable of earning superior profits and surviving the economic cycles that so directly affect the construction industry. In addition to these general benefits, regular and effective financial analysis allows the firm to:

- Understand and control its growth, as well as the associated increase in operational complexities.
- Compete in an ever more difficult business environment by controlling costs, refining its financial structure, and operating at the lowest costs possible.
- Establish meaningful financial goals and guidelines as important components in decision making.
- Understand and present its financial strengths and weaknesses to external creditors in the most favorable light possible.
- Detect and respond to specific operational shortfalls in such areas as project management, financial management, and marketing.

Construction firms depend heavily on a number of external parties who have a vested interest in their financial strength. Given the high levels of capital often required, contractors must establish and maintain good relationships with various lending institutions. For most firms, the availability of bonding through surety firms is a crucial factor in acquiring the larger and more lucrative projects. Building firms must also establish short term credit agreements with material suppliers and subcontractors. Finally, the response of public and private owners in need of construction services is clearly going to be affected by the ability of a contractor to perform without encountering financial difficulties during the building process.

External parties are all rightfully concerned about the financial strength of the contractor, particularly in light of the high industry failure rate. Their primary means of evaluating a firm's financial strength and ability to perform is through analysis of its financial statements. In many cases, the analysis made by these external parties actually determines the firm's ability to continue operations, borrow funds, and pursue larger sales revenues. For this reason, the contractor must become aware of financial analysis techniques and be fully prepared to present the firm's position in the most favorable light possible. With expert interpretation, he can prevent erroneous or misdirected conclusions that can have a serious impact on the firm's ability to operate in the future.

The Contractor's Financial Structure

A number of unique factors influence the financial structure of construction firms. Interpreting a contractor's financial data can be somewhat difficult as a result. Some of the industry's special characteristics cause construction firms to operate under less than ideal circumstances, sometimes with an uncomfortable level of risk. While all of these factors do not necessarily affect every firm, each should be considered and addressed as thoroughly as possible.

While manufacturing firms may produce the same general products or services many times over for the same price each time, contractors provide very diverse services in the construction of only a few very diverse buildings, and for a very large price each time. The nature of the business breeds complexity, a great potential for conflict, and a high dependence on the success of each project. The systems required to control and report the firm's performance in this environment must, therefore, be finely tuned and provide management with valuable and timely information.

Debt

With the exception of a few very large national and international publicly-held contracting firms, most construction companies are closely-held business entities, the activities of which are limited to a small geographic area. As a result, they cannot raise equity capital by selling stock, and must depend on both long and short term debt. The impact of debt on a firm's financial strength is significant, particularly when sales revenues are low and the ability to service the debt is diminished. Because of the high risk involved in contracting, lenders are reluctant to commit substantial funds to the firm on a long term basis, opting instead for short term notes and lines of credit. Unfortunately, because short term liabilities become due within one year, the firm is placed in a position of still higher risk.

Receivables

Contractors must often commit a high percentage of their assets to receivables. As the projects become larger, monthly draws against the total contract amount also become quite large. The delay of just a few days in receipt of a single sizable draw, much less several draws on different projects, can create immediate cash problems. To make matters worse, the retainage often found in commercial contracting typically exceeds the real profit on a project; the result is that the firm's cash reserves are slowly drained to cover operating expenses. Avoiding cash flow problems due to slow receivables requires contractual language that clearly outlines the owner's payment obligations and the forms of recourse available to the contractor. Lenders should be apprised of these terms before construction begins in order to avoid subsequent delays or disputes. In addition, an ongoing system to follow up on receivables is crucial in avoiding delayed payments.

Fixed Assets

Most contractors must invest relatively large sums in fixed assets in order to operate. In addition to the normal tenant improvements and office equipment required for most businesses, contractors also have to purchase trucks and tools, and may have to invest in heavy equipment. A shop to house the equipment and any material inventory is also a necessity. Because of the heavy borrowing involved with such assets, the firm is highly exposed during recessionary periods when sales revenues are reduced. Consequently, each proposed expenditure for fixed assets must be carefully evaluated.

Financial Analysis Considerations

While ratio and trend analysis are extremely helpful in evaluating a firm's financial strength, several influencing factors must also be considered in the process. It is all too easy to draw the wrong conclusions from either incomplete or improperly prepared data. A full understanding of the ratio and trend information generated through financial analysis requires:

- Analysis of comparative data from successive periods of activity, both for the individual firm and the industry as a whole.
- Analysis of various events influencing the firm's performance during the designated period.
- Evaluation of the overall performance and unique characteristics of the industry in general.

A number of factors complicate efforts to compare and interpret the financial data collected by general contracting firms. The most significant of these factors are described in the following paragraphs, although a number of lesser issues may also come into play.

Accounting Practices

As explained in Chapter 2, several different accounting methods are available to contractors. The figures represented in financial statements prepared using *cash versus accrual* and *percentage-of-complete versus completed contract* accounting methods can be substantially different. While the *accrual basis* and *percentage-of-complete* methods are most meaningful for managerial accounting reports and financial analysis, the *cash basis* and *completed contract* methods may be highly beneficial for tax reporting purposes. The impact of these practices on a contractor's financial statement must be understood by the analyst.

Asset Ownership

The treatment of fixed assets plays a major role in the makeup of the firm's financial structure. Some business owners may opt to purchase the firm's office space and other fixed assets personally, renting them to the firm at rates somewhat higher or lower than the market may dictate. While the owner's tax and financial goals may be best served by such practices, the true profitability and financial strength of the firm may be distorted. Likewise, many firms lease fixed assets in lieu of direct purchase. While such facts are typically disclosed in the financial statements, this leaves a potential for misinterpretation of the firm's position regarding debt, equity, operating expenses, and profitability.

Company Structure

Large companies with several profit centers are more difficult to evaluate than small or medium-sized companies operating as a single profit center. The performance of individual divisions within the larger company is not necessarily revealed on the firm's consolidated financial statements. In such cases, effective analysis requires an evaluation of each division for its individual financial performance and contribution to the company as a whole.

Seasonal Conditions

Contractors operating in areas that are seriously affected by unfavorable seasonal weather conditions pose a special challenge to the financial analyst. For instance, a firm in the northern part of the country may appear financially healthier in November, after the best construction months, and less stable in March, after the high cost and low productivity winter months. Likewise, two or more consecutive winters of unusually harsh weather may have a substantial negative effect, compared to the positive effect of consecutive mild winters. In this same vein, contractors whose work involves construction in resort areas are likely to experience similar fluctuations in activity.

Data Distortion

It is common to find that some ratios and trends appear strong, while others are weak. This distortion may leave some doubt as to the true financial strength of the company, but an in-depth cause-and-effect study usually reveals the facts behind the distortion. One very sophisticated approach to this dilemma is known as *multiple discriminant analysis*. This process involves reducing the firm's ratios to indexes for comparison with indexes of ratios of similar firms. By analyzing the movement of various firms' indexes over time, it is possible to pinpoint the cause of data distortions and to clearly understand the current strength of the individual firm.

One significant distortion is in the level of different balance sheets and income statement accounts at the end of the period, versus the average during the course of the period. For instance, a firm with net profit before taxes of $100,000 in 1988 and a net worth of $300,000 at the beginning of 1988 would report a return on equity of $100,000/$300,000, or 33%. However, the *average* net worth of the firm during the course of the period was probably closer to $350,000, assuming the earnings of $100,000 were spread evenly over the year. The actual return would then be computed as $100,000/$350,000, or 28%.

Similar distortions often occur in several other ratios that may be more appropriately computed using average account figures. Another example is the contractor who has $4,000,000 in annual direct costs and average open payables of $450,000. If the firm manages to close the year with payables of $250,000, the payable turns ratio would indicate much better performance in meeting payable obligations. Using the average payables, we get $4,000,000/$450,000 = 8.9 turns per year, or an average of 41 days from invoice receipt to payment. However, the end of period payables gives us $4,000,000/$250,000 = 16 turns per year, or an average of 23 days for payment.

In these examples, it is easy to recognize the value of using average account balances in computing a number of the typical ratios. However, while this method may offer more accurate ongoing performance reporting for the individual firm, the comparative industry data reported by such firms as Robert Morris Associates, is based on year-end data. Therefore, a contractor comparing his ratios to those provided by major accounting firms for similar businesses must also use year-end ratios.

Magnitude
Financial analysis can easily be misleading if the magnitude of the numbers is not considered. For instance, a small firm just starting in the business may well see its revenues expand from $500,000 in the first year to $2,000,000 the following year, or a 400% increase. However, it would indeed be difficult for a firm doing $20,000,000 in sales to increase 400% in one year. The increase in return on equity is likely to be much higher for the small firm during this period, for its net worth would no doubt be very small compared to its earnings if the firm is effectively managed.

Officer's Compensation
The owners of construction firms may opt to draw a salary higher or lower than is common in the industry in order to accomplish their own financial or tax goals. In a small or medium-sized firm, this decision may have a significant impact on the firm's profitability and working capital. Some owners borrow funds from the firm, reflecting these amounts as accounts receivable despite the fact that they have no intention of paying them back in the near term. While the solvency of the firm is certainly affected by this practice, the financial statements do not necessarily reflect its impact.

Industry Changes
The ideal financial structure allows a construction firm to be dynamic, changing from time to time in response to a number of factors. What was optimum in the past may not be today, for the business environment is in a constant state of flux. Just a few factors that may dictate desirable changes in the financial structure of a company include new accounting standards, changing tax laws, economic conditions, and work backlog. Depending on the firm's response to these factors, there may be valid reasons for ratios or trends to vary somewhat from industry standards.

Internal Manipulation
Managers often have some flexibility in manipulating their firm's financial data to reflect a stronger position on financial statements. For instance, a contractor reporting on the accrual basis percentage-of-complete method may press his subcontractors to submit slightly inflated bills just prior to the closeout of a period, which, in turn, establishes a higher level of earned revenues. The estimated job cost figures for a given project may be adjusted to reflect a higher or lower gross margin to be earned, creating or diluting profit earned to date. A number of other such practices may substantially distort the true performance of the firm.

Data Classification

Statistical data must be collected and properly classified in order to allow effective comparative analysis. Unfortunately, the chart of accounts developed by individual contractors often fails to include crucial data in accordance with the procedures and classifications recognized in the industry. Direct comparison, either to other firms or to data provided by such organizations as Robert Morris Associates, then becomes quite difficult, if not impossible.

Using Financial Ratios

A ratio is simply a mathematical expression of the relationship of one piece of data to another. Ratios are used to evaluate financial statements because they allow numbers to be summarized and related in a manner that is easy to understand. Relationships of key financial data can point up the conditions and trends of a firm's financial structure.

Individual ratios may be considered alone to express a relationship between two items at a certain point in time, or as a series of ratios from two or more periods. Valuable insights can be obtained by comparing these ratios with those of other similar companies during the same period. There are several sources of such comparative financial data, including Dunn and Bradstreet and Robert Morris Associates. These services publish manuals that outline the make-up of the major financial statement classifications (assets, liabilities, net worth and income data) as well as financial ratios for businesses in the following categories: manufacturing, wholesale, service, retail, and contracting. Managers in almost any business can use the information available through these services to compare their performance with the lower, median and upper quartile performance of a large cross section of similar businesses.

Ratios are also used in trend analysis. In this case, ratios from two or more periods are compared, and the result shows how a specific financial relationship has changed over time. For instance, through ratio analysis we may learn that a firm has reduced its average period from invoice receipt to payment from 30 days in 1987 to 20 days in 1988. While this is a positive trend, we might also learn that the same firm collected its receivables in 15 days in 1987, versus 25 days in 1988, clearly a negative trend. If left unattended, the combination of these two trends would soon have a serious impact on the firm's cash reserves.

Financial ratios may be computed using two balance sheet accounts, two income statement accounts, or one balance sheet and one income statement account. In some cases, financial data may also be combined with non-financial data, such as the number of employees or days in a period.

Ratios should be expressed in a form that is most easily understood. There are three standard approaches:

1. As a percentage, e.g., the firm earned net profits of 4% on sales of $5,000,000.
2. As x to 1, e.g., the firm's ratio of current assets to current liabilities is 2 to 1.
3. As x per y, e.g., the firm averages $1,000,000 in annual sales per field supervisor.

A number of standard financial ratios are described in the following sections. Each of these ratios represents an important relationship between the items in a construction firm's financial reports. These ratios provide guidelines that are applicable to the construction industry. Thus, managers are able to compare their financial results with those of similar firms. Individual firms may benefit by exploring other commonly used ratios, or developing their own ratios for tracking performance in very specialized areas.

The ratios can be organized into the following four categories:
1. *Liquidity ratios*, that reflect the firm's ability to meet its financial obligations as they become due.
2. *Leverage ratios*, that evaluate the degree to which the firm has leveraged its equity, and the associated risk to creditors in the event the firm encounters financial difficulties.
3. *Profitability ratios*, that measure the firm's success in using its resources to earn favorable returns.
4. *Cost ratios*, that compare total or specific operating expenses to total sales.

These ratios are defined in the following sections, using the hypothetical construction firm, Eastway Contractors, as an example. The firm is representative of many building firms that proceed through several stages of financial and managerial development before becoming highly effective and profitable. Through several years of skilled analysis of its financial ratios and trends, and effective response to the conclusions drawn, the firm has developed a position of superior performance and strength—particularly in 1987. The section on *Trend Analysis* which appears later in this chapter explains the strategies undertaken by Eastway Contractors to reach this point.

The examples in the following sections focus on the firm's 1987 income statement and balance sheets, introduced in Chapter 2. The data from those statements is consolidated in Figures 3.1 through 3.3, which reflects a spreadsheet format used in trend analysis.

The median figures shown are derived from the *1987 Annual Statement Studies* published by Robert Morris Associates. The specific reference is *Standard Industrial Classification (SIC) # 1541 for Contractors — Commercial Construction*, shown in Figure 3.4. The specific data is drawn from the column for sales revenues of 1–10 million dollars.

Liquidity Ratios

Liquidity refers to a firm's ability to respond quickly to current debt, obligations, and investment opportunities. Without a proper level of liquidity, the firm is likely to incur cash flow problems in meeting payables. Over time, a lack of liquidity may prevent the firm from borrowing needed funds, increasing its sales volume, or establishing new credit facilities with vendors.

Current Ratio: The level of liquidity is directly linked to the firm's working capital position at any given time. *Working capital* is the excess of current assets over current liabilities and represents the net cash available for operations. The level of working capital is crucial to the firm's financial stability, for it is often necessary in the contracting business to pay vendor invoices prior to receiving payments from owners. In addition, retainage is sometimes withheld from construction draws, preventing the firm from realizing its profit, and possibly a portion of its

EASTWAY CONTRACTORS, INC.
INCOME STATEMENT
TREND ANALYSIS

	PERCENTAGE OF TOTAL				FINANCIAL DATA		
	MED. %	GOAL	1987	1986	1985	1987	1986	1985
CONTRACT REVENUES	100.0%	100.0%	100.0%	100.0%	100.0%	6,000,000	4,000,000	3,000,000
DIRECT COSTS								
Labor	----	5.0%	6.3%	9.6%	13.7%	375,000	385,000	410,000
Material	----	13.0%	13.2%	21.5%	32.3%	790,000	860,000	970,000
Subcontractor	----	65.0%	64.6%	51.3%	33.3%	3,875,000	2,050,000	1,000,000
Equipment	----	1.0%	0.8%	2.5%	5.0%	50,000	100,000	150,000
Miscellaneous	----	1.0%	0.8%	1.5%	2.7%	50,000	60,000	80,000
TOTAL DIRECT COSTS	85.2%	85.0%	85.7%	86.4%	87.0%	5,140,000	3,455,000	2,610,000
GROSS PROFIT	14.8%	15.0%	14.3%	13.6%	13.0%	860,000	545,000	390,000
OPERATING EXPENSES								
INDIRECT COSTS								
Field Employees	----	0.6%	0.6%	0.8%	0.9%	33,000	30,000	28,000
Field Equipment	----	0.3%	0.3%	0.5%	0.8%	16,000	18,000	23,000
Field Insurance	----	0.4%	0.4%	0.5%	0.5%	26,000	21,000	16,000
Field Vehicles	----	0.4%	0.4%	0.8%	1.2%	23,000	30,000	35,000
Shop Maintenance	----	0.1%	0.1%	0.2%	0.3%	4,000	6,000	8,000
Warranty	----	0.3%	0.3%	0.6%	0.8%	18,000	23,000	25,000
Miscellaneous Indirect	----	0.2%	0.2%	0.2%	0.2%	10,000	7,000	5,000
TOTAL INDIRECT COSTS	----	2.3%	2.2%	3.4%	4.7%	130,000	135,000	140,000
GENERAL & ADMINISTRATIVE EXPENSES								
Business Insurance	----	0.1%	0.1%	0.1%	0.1%	8,000	5,000	4,000
Occupancy	----	0.8%	0.8%	0.9%	0.8%	45,000	35,000	25,000
Office Employees	----	2.5%	2.6%	2.9%	2.7%	155,000	114,000	82,000
Office Equipment	----	0.2%	0.2%	0.2%	0.1%	9,000	6,000	4,000
Office Supplies	----	0.2%	0.2%	0.3%	0.4%	13,000	11,000	12,000
Office Vehicles	----	0.2%	0.2%	0.2%	0.2%	13,000	8,000	7,000
Officer Compensation	----	2.3%	2.7%	2.3%	3.3%	160,000	90,000	100,000
Professional Fees	----	0.2%	0.2%	0.2%	0.1%	12,000	6,000	3,000
Sales & Marketing	----	0.7%	0.7%	0.4%	0.1%	40,000	15,000	4,000
Travel & Entertainment	----	0.2%	0.2%	0.2%	0.2%	10,000	7,000	6,000
Other General & Admin.	----	0.3%	0.3%	0.3%	0.3%	15,000	11,000	9,000
TOTAL GENERAL & ADMIN.	----	7.7%	8.0%	7.7%	8.5%	480,000	308,000	256,000
TOTAL OPERATING EXPENSES	12.3%	10.0%	10.2%	11.1%	13.2%	610,000	443,000	396,000
OPERATING PROFIT(LOSS)	2.6%	5.0%	4.1%	2.5%	-0.2%	250,000	102,000	(6,000)
ALL OTHER INCOME(EXPENSES)-NET	0.3%	0.0%	0.2%	0.1%	-0.2%	10,000	2,000	(5,000)
PROFIT(LOSS) BEFORE TAXES	2.9%	5.0%	4.0%	2.5%	0.0%	260,000	104,000	(11,000)

Figure 3.1

EASTWAY CONTRACTORS, INC.
BALANCE SHEET
TREND ANALYSIS

	PERCENTAGE OF TOTAL				FINANCIAL DATA		
	MED. %	GOAL	1987	1986	1985	1987	1986	1985
ASSETS								
CURRENT ASSETS								
Cash and Equivalents	18.5%	30.0%	29.9%	20.5%	11.9%	295,000	160,000	85,000
A/R-Progress Billings	39.9%	40.0%	40.6%	41.0%	35.9%	400,000	320,000	257,000
A/R-Current Retention	6.9%	6.0%	8.1%	9.0%	9.1%	80,000	70,000	65,000
Inventory	2.7%	1.0%	1.0%	1.9%	3.5%	10,000	15,000	25,000
Costs & Earnings Exceeding Bills	5.5%	5.0%	5.1%	5.8%	7.7%	50,000	45,000	55,000
All Other Current	4.2%	2.0%	2.5%	1.3%	1.1%	25,000	10,000	8,000
TOTAL CURRENT ASSETS	77.7%	84.0%	87.3%	79.5%	69.2%	860,000	620,000	495,000
NON-CURRENT ASSETS								
Fixed Assets (net)	14.3%	10.0%	8.6%	14.1%	19.6%	85,000	110,000	140,000
Joint Ventures & Investments	1.6%	0.0%	0.0%	1.9%	7.0%	0	15,000	50,000
Intangibles	0.2%	0.0%	0.0%	0.0%	0.0%	0	0	0
All Other Non-Current	6.2%	6.0%	4.1%	4.5%	4.2%	40,000	35,000	30,000
TOTAL NON-CURRENT ASSETS	22.3%	16.0%	12.7%	20.5%	30.8%	125,000	160,000	220,000
TOTAL ASSETS	100.0%	100.0%	100.0%	100.0%	100.0%	985,000	780,000	715,000
LIABILITIES AND NET WORTH								
CURRENT LIABILITIES								
Notes Payable-Short Term	6.2%	5.0%	3.0%	5.1%	7.0%	30,000	40,000	50,000
A/P-Trade	30.2%	30.0%	26.9%	35.3%	39.9%	265,000	275,000	285,000
A/P-Retention	1.8%	2.0%	2.5%	3.2%	2.8%	25,000	25,000	20,000
Bills Exceeding Costs & Earnings	5.4%	8.0%	6.1%	6.4%	5.6%	60,000	50,000	40,000
Income Taxes Payable	1.8%	2.0%	2.5%	1.3%	0.0%	25,000	10,000	0
Current Maturity LTD	2.5%	1.0%	1.5%	2.6%	4.2%	15,000	20,000	30,000
All Other Current	6.1%	5.0%	2.5%	2.9%	2.8%	25,000	23,000	20,000
TOTAL CURRENT LIABILITIES	54.0%	53.0%	45.2%	56.8%	62.2%	445,000	443,000	445,000
NON-CURRENT LIABILITIES								
Long Term Debt (LTD)	7.5%	4.0%	3.0%	5.1%	8.4%	30,000	40,000	60,000
Deferred Taxes	2.2%	2.0%	2.5%	0.6%	0.0%	25,000	5,000	0
All Other Non-Current	0.9%	1.0%	1.5%	1.5%	1.4%	15,000	12,000	10,000
TOTAL NON-CURRENT LIAB.	10.6%	7.0%	7.1%	7.3%	9.8%	70,000	57,000	70,000
NET WORTH (NW)	35.4%	40.0%	47.7%	35.9%	28.0%	470,000	280,000	200,000
TOTAL LIABILITIES & NW	100.0%	100.0%	100.0%	100.0%	100.0%	985,000	780,000	715,000
OTHER KEY DATA								
Average Work-in-Process						3,500,000	2,000,000	1,300,000
Working Capital						415,000	177,000	50,000
Interest Expense						8,000	10,000	15,000
Depreciation & Amort. Expense						20,000	30,000	40,000

Figure 3.2

EASTWAY CONTRACTORS, INC.
FINANCIAL RATIOS
TREND ANALYSIS

	Median	Upper Quartile	1987	1986	1985	FORMULA
LIQUIDITY RATIOS						
Current Ratio	1.4	1.8	1.9	1.4	1.1	Current Assets/ Current Liabiliites
Receivable Turns	7.0	10.8	12.5	10.3	9.3	Revenues/Receivables
Payable Turns	9.2	14.6	17.7	11.5	8.6	Direct Costs/Payables
Working Capital Turns	14.3	8.5	14.5	22.6	60.0	Revenues/Working Capital
Working Capital to Average W.I.P.	----	----	11.9%	8.9%	3.8%	Working Capital/Work-in-Process
DEBT RATIOS						
Earnings Coverage Over Interest	6.7	17.8	31.3	10.2	-0.4	Operating Profit/ Interest Expense
Cash Flow Coverage Over Current Portion LTD	4.7	9.9	18.7	6.7	1.0	(Net Profit + Deprec. + Amort.)/Current Maturity LTD
Portion of Equity Invested in Fixed Assets	0.3	0.1	0.2	0.4	0.7	Fixed Assets/Net Worth
Debt to Net Worth	1.9	1.1	1.1	1.8	2.6	Total Liabilities/Net Worth
OPERATING RATIOS						
Gross Profit	14.8%	----	14.3%	13.6%	13.0%	Gross Profit/Revenue
Net Profit	2.9%	----	4.3%	2.6%	-0.4%	Net Profit/Revenue
Return on Equity	21.9%	47.6%	55.3%	37.1%	-5.5%	Profit Before Tax/Net Worth
Return on Assets	7.4%	15.6%	26.4%	13.3%	-1.5%	Profit Before Tax/Total Assets
COST RATIOS						
Direct Costs to Sales	85.1%	----	85.7%	86.4%	87.0%	Direct Costs/Revenue
Indirect Costs to Sales	----	----	2.2%	3.4%	4.7%	Indirect Costs/Revenue
General & Administrative Costs to Sales	----	----	8.0%	7.7%	8.5%	General & Admin. Costs/Revenue
Total Operating Expense to Sales	12.3%	----	10.2%	11.1%	13.2%	(Indirect Costs + G & A)/Revenue

Figure 3.3

CONTRACTORS—COMMERCIAL CONSTRUCTION SIC# 1541 (42)

Current Data					Type of Statement	Comparative Historical Data				
13	300	178	45	536	Unqualified			487	440	536
2	19	13	6	40	Qualified			52	39	40
60	327	53	2	442	Reviewed	DATA NOT AVAILABLE		303	334	442
41	96	11	1	149	Compiled			114	116	149
11	45	15	3	74	Other			77	82	74
545(6/30-9/30/86)			696(10/1/86-3/31/87)			6/30/82-3/31/83	6/30/83-3/31/84	6/30/84-3/31/85	6/30/85-3/31/86	6/30/86-3/31/87
0-1MM	1-10MM	10-50MM	50 & OVER	ALL	CONTRACT REVENUE SIZE	ALL	ALL	ALL	ALL	ALL
127	787	270	57	1241	NUMBER OF STATEMENTS	864	853	1033	1011	1241
%	%	%	%	%	**ASSETS**	%	%	%	%	%
19.8	18.5	17.4	14.3	18.2	Cash & Equivalents	20.8	19.9	16.3	18.1	18.2
30.4	39.9	43.5	40.5	39.7	A/R - Progress Billings	35.2	36.5	39.1	39.1	39.7
4.6	6.9	10.5	14.2	7.8	A/R - Current Retention	9.6	8.0	8.1	8.0	7.8
4.5	2.7	1.9	1.1	2.6	Inventory	2.3	2.7	3.3	2.4	2.6
3.6	5.5	5.0	4.6	5.1	Costs & Est. Earnings in Excess of Billings	3.9	4.6	4.8	5.0	5.1
5.3	4.2	4.3	4.8	4.4	All Other Current	4.0	3.7	4.2	3.8	4.4
68.2	77.7	82.6	79.4	77.9	Total Current	75.9	75.4	75.7	76.3	77.9
19.4	14.3	9.8	12.3	13.7	Fixed Assets (net)	13.5	13.9	14.3	14.9	13.7
2.9	1.6	1.9	2.9	1.9	Joint Ventures & Investments	2.7	2.5	2.3	1.6	1.9
.7	.2	.4	.4	.3	Intangibles (net)	.2	.3	.3	.3	.3
8.8	6.2	5.2	4.9	6.2	All Other Non-Current	7.7	7.8	7.4	6.9	6.2
100.0	100.0	100.0	100.0	100.0	Total	100.0	100.0	100.0	100.0	100.0
					LIABILITIES					
9.3	6.2	4.3	2.3	5.9	Notes Payable - Short Term	5.2	6.0	7.2	6.1	5.9
19.7	30.2	35.0	35.7	30.4	A/P - Trade	28.6	29.5	29.6	30.4	30.4
.6	1.8	6.0	7.3	2.8	A/P - Retention	3.5	2.7	3.2	3.0	2.8
2.8	5.4	7.0	7.9	5.6	Billings in Excess of Costs & Est. Earnings	5.7	5.4	5.6	5.7	5.6
1.4	1.8	2.0	1.4	1.8	Income Taxes Payable	—	—	1.5	1.6	1.8
3.5	2.5	2.0	1.4	2.4	Cur. Mat. - L/T/D	2.4	2.0	2.6	2.5	2.4
8.5	6.1	7.2	7.9	6.7	All Other Current	9.1	8.7	7.3	6.4	6.7
45.8	53.9	63.5	63.9	55.6	Total Current	54.3	54.4	56.9	55.8	55.6
13.0	7.5	5.7	8.2	7.7	Long Term Debt	6.8	7.6	7.8	7.0	7.7
1.1	2.2	2.9	4.0	2.3	Deferred Taxes	—	—	1.9	2.2	2.3
1.0	.9	.9	.7	.9	All Other Non-Current	3.0	2.7	1.0	1.2	.9
39.2	35.4	27.0	23.2	33.4	Net Worth	35.9	35.4	32.5	33.8	33.4
100.0	100.0	100.0	100.0	100.0	Total Liabilities & Net Worth	100.0	100.0	100.0	100.0	100.0
					INCOME DATA					
100.0	100.0	100.0	100.0	100.0	Contract Revenues	100.0	100.0	100.0	100.0	100.0
27.3	14.8	8.9	7.3	14.5	Gross Profit	13.4	13.7	14.8	14.6	14.5
24.3	12.3	7.1	5.9	12.1	Operating Expenses	12.4	13.2	12.6	12.0	12.1
3.1	2.6	1.7	1.3	2.4	Operating Profit	1.0	.5	2.1	2.6	2.4
.3	-.3	-.3	.0	-.2	All Other Expenses (net)	-.5	-.5	-.2	-.3	-.2
2.8	2.9	2.0	1.3	2.6	Profit Before Taxes	1.5	1.0	2.3	2.9	2.6
					RATIOS					
2.5	1.8	1.5	1.4	1.7	Current	1.8	1.8	1.7	1.7	1.7
1.4	1.4	1.3	1.3	1.4		1.4	1.4	1.3	1.3	1.4
1.0	1.2	1.2	1.1	1.2		1.2	1.1	1.1	1.1	1.2
3.8	2.1	1.7	1.6	2.1	Receivables/Payables	1.9	2.0	2.1	2.0	2.1
(125) 2.0	(785) 1.5	(269) 1.3	1.2	(1236) 1.4		(861) 1.4	(844) 1.4	(1024) 1.4	(999) 1.4	(1236) 1.4
1.1	1.1	1.1	1.0	1.1		1.1	1.0	1.1	1.1	1.1
26 14.3	34 10.8	42 8.7	47 7.7	35 10.3	Revenues/Receivables	36 10.1	35 10.3	38 9.7	37 9.8	35 10.3
46 7.9	52 7.0	57 6.4	58 6.3	54 6.8		53 6.9	53 6.9	57 6.4	54 6.8	54 6.8
72 5.1	72 5.1	72 5.1	70 5.2	72 5.1		73 5.0	72 5.1	78 4.7	72 5.1	72 5.1
12 30.3	25 14.6	32 11.5	33 11.2	25 14.4	Cost of Revenues/Payables	26 14.3	25 14.5	26 14.2	26 13.9	25 14.4
33 10.9	40 9.2	46 8.0	50 7.3	41 9.0		42 8.7	42 8.6	43 8.4	42 8.6	41 9.0
65 5.6	57 6.4	63 5.8	62 5.9	59 6.2		61 6.0	61 6.0	63 5.8	59 6.2	59 6.2
5.9	8.5	12.2	14.1	9.1	Revenues/Working Capital	8.3	7.8	9.4	9.7	9.1
12.3	14.3	21.0	22.7	15.8		15.3	15.8	18.0	17.0	15.8
-248.5	28.2	35.3	39.4	32.2		33.3	41.6	44.6	38.2	32.2
8.8	17.8	24.0	11.9	18.3	EBIT/Interest	16.2	12.0	13.8	17.3	18.3
(92) 3.6	(639) 6.7	(229) 8.9	(42) 4.5	(1002) 6.6		(647) 4.0	(647) 3.3	(774) 4.3	(788) 5.5	(1002) 6.6
.7	2.1	2.2	1.9	2.0		1.0	.0	1.4	2.0	2.0
5.3	9.9	11.4	8.5	9.6	Net Profit + Depr., Dep., Amort./Cur. Mat. L/T/D	11.2	9.5	7.7	10.3	9.6
(42) 2.7	(459) 4.7	(160) 4.3	(32) 3.4	(693) 4.3		(452) 3.3	(415) 3.6	(515) 3.1	(529) 3.8	(693) 4.3
.8	1.7	1.7	1.2	1.6		1.0	1.0	.9	1.5	1.6
.1	.1	.2	.2	.2	Fixed/Worth	.1	.1	.2	.2	.2
.4	.3	.3	.3	.3		.3	.3	.3	.3	.3
1.2	.6	.5	.8	.6		.6	.6	.7	.7	.6
.6	1.1	1.8	2.5	1.2	Debt/Worth	1.1	1.1	1.3	1.2	1.2
1.7	1.9	3.1	4.1	2.2		2.0	2.0	2.4	2.2	2.2
4.6	3.2	4.8	5.7	4.0		3.3	3.6	4.4	4.0	4.0
44.5	47.6	45.9	43.0	46.8	% Profit Before Taxes/Tangible Net Worth	34.9	30.0	41.2	45.4	46.8
(117) 19.0	(760) 21.9	(267) 26.5	20.4	(1201) 22.7		(845) 15.3	(836) 12.5	(996) 18.9	(984) 22.7	(1201) 22.7
.1	6.6	9.4	10.1	6.8		.8	-2.2	4.1	6.7	6.8
18.0	15.6	11.1	7.9	14.5	% Profit Before Taxes/Total Assets	11.4	10.1	12.0	13.8	14.5
6.7	7.4	6.0	4.0	6.7		5.1	4.0	5.5	6.8	6.7
-.6	2.1	2.4	2.1	2.0		.1	-1.2	.9	2.0	2.0
.7	.5	.3	.3	.4	% Depr., Dep., Amort./Revenues	.4	.4	.5	.4	.4
(107) 1.8	(721) .9	(254) .4	(50) .5	(1132) .8		(801) .8	(767) .9	(902) .8	(901) .8	(1132) .8
3.4	1.6	.8	.8	1.5		1.6	1.7	1.6	1.7	1.5
4.5	1.6	.9		1.5	% Officers' Comp./Revenues	1.6	1.5	1.6	1.5	1.5
(68) 5.9	(397) 2.8	(97) 1.4		(571) 2.8		(376) 2.8	(391) 2.7	(447) 2.6	(447) 2.5	(571) 2.8
8.2	4.4	2.8		4.4		4.5	5.6	4.4	4.6	
75599M	3303278M	5728383M	7809321M	16916581M	Contract Revenues ($)	18188584M	10678876M	18584853M	14388778M	16916581M
43149M	1267451M	1930530M	3012344M	6253474M	Total Assets ($)	5365296M	4285788M	6969729M	5260282M	6253474M

© Robert Morris Associates 1987

M = $thousand   MM = $million
See Pages 1 through 13 for Explanation of Ratios and Data

Figure 3.4

operating expenses, until the project is complete. Finally, sureties consider working capital to be of utmost importance in establishing the firm's bondable level of work-in-process. The following ratio reflects the firm's current liquidity.

$$\text{Current Ratio} = \frac{\text{Current Assets}}{\text{Current Liabilities}}$$

The current ratio is the most commonly used indicator of the firm's current level of solvency. It represents the coverage of current assets over current liabilities. The higher the ratio, the greater the firm's ability to meet its current obligations. A ratio of anything less than 1 indicates a cash flow problem in the short term. Eastway's current ratio for 1987 is derived as follows:

$$\frac{\text{Current Assets}}{\text{Current Liabilities}} \quad \frac{\$860,000}{\$445,000} = 1.9$$

This ratio occurs near the upper quartile of 1.8, and reflects the firm's strong current liquidity position. The difference between current assets and current liabilities represents *working capital*. In the case of Eastway, this figure is derived from the following calculation:

$$\$860,000 - \$445,000 = \$415,000$$

If the firm's total sales of $6,000,000 are evenly distributed throughout the year, $415,000 should be quite a sufficient amount of working capital for Eastway.

Receivable Turns: The ratio for receivable turns measures the number of times that receivables turn over during an operating period. A higher number of turns reflects a shorter time between the date that a bill is issued, and the date when the cash is collected. A firm with a low receivables turnover rate may not have an effective receivables follow-up plan, poor contractual protection, disputes with customers, or a number of other possible problems. In such cases, the use of an average receivables figure in lieu of a year-end figure would be more indicative of the firm's overall receivables performance.

$$\text{Receivable Turns} = \frac{\text{Revenue}}{\text{Receivables}}$$

Receivable turnover for Eastway is computed as follows:

$$\frac{\text{Revenue}}{\text{Receivables}} \quad \frac{\$6,000,000}{\$480,000} = 12.5$$

This ratio indicates that the firm turned over its receivables 12.5 times in 1987, or slightly more often than the upper quartile figure, which stands at 10.8. By simply dividing the number of turns into the number of days in a year, it is possible to determine the average time—in days required—from the issuance of bills to cash collection. In the case of Eastway Contractors, this figure would be:

$$\frac{\text{Days In Year}}{\text{Receivable Turns}} \quad \frac{365}{12.5} = 29 \text{ days}$$

Payable Turns: The Payable Turns ratio reflects the frequency of payable turnovers during the period. A higher number indicates more frequent payable turns and, therefore, a reduced time from the point of invoice receipt to the point of payment. A low turnover rate could be caused by insufficient working capital, special credit arrangements with vendors, or perhaps disputed bills. Again, an *average* payables figure would be more meaningful for use in this ratio than an *end of period* payables figure.

$$\text{Payable Turns} = \frac{\text{Direct Costs}}{\text{Payables}}$$

The payable turnover rate for Eastway is computed as:

$$\frac{\text{Direct Costs}}{\text{Payables}} \quad \frac{\$5,140,000}{\$290,000} = 17.7$$

The firm turned over its payables 17.7 times during the year, considerably better than the upper quartile figure of 14.6. Dividing 365 (days per year) by the number of payable turns yields the average number of days from invoice receipt to payment. For Eastway, this would be:

$$\frac{\text{Days In Year}}{\text{Payable Turns}} \quad \frac{365}{17.7} = 20.6 \text{ days}$$

An average of 20 days to meet payable obligations would be considered quite ample in the construction industry. All subcontractors and suppliers want to be paid in the shortest time possible, and are often willing to offer a discount incentive for prompt payment. The level of quality and general service is also likely to be greater from vendors who feel assured of rapid payment.

Working Capital Turns: The ability of any firm to produce the maximum return relative to resources is an important determinant of profitability. Inasmuch as working capital is a resource, and the level of sales plays a major role in profits, the relationship of working capital turnover to sales is significant. In other words, a firm that is able to produce sales equal to 20 times the level of working capital, as compared to 5 times the working capital, is using working capital very effectively. On the other hand, a firm with an excessive number of working capital turns may simply be undercapitalized, and at the risk of receivable and payable problems.

$$\text{Working Capital Turns} = \frac{\text{Revenue}}{\text{Working Capital}}$$

Working capital turns for Eastway can be derived as:

$$\frac{\text{Revenue}}{\text{Working Capital}} \quad \frac{\$6,000,000}{\$415,000} = 14.5$$

A turnover of 14.5 times is quite strong when compared to the upper quartile of 8.5. Interestingly, the upper quartile in this case has a typical turnover rate for working capital of fewer times than the median. This favorable comparison is due to the tendency of the more established and well-capitalized firms to maintain a much higher level of working capital than those firms that are less financially stable. A firm with an extraordinarily high number of working capital turns is probably undercapitalized, and runs the risk of cash flow shortages. The ratio of working capital to average work-in-process is explained next.

Working Capital to Average Work-in-Process: The sales volume comfortably managed by a contracting firm is often a direct function of the level of working capital available. Bonding firms are particularly sensitive to this fact, and typically look for a level of working capital somewhere between 7% and 10% of work-in-process. The reason is simple: as sales increase, greater demands are placed on the firm in meeting such current obligations as accounts payable, wages, and operating expenses. This relationship for Eastway is calculated as follows:

$$\frac{\text{Working Capital}}{\text{Average Work-in-Process}} \quad \frac{\$\,415,000}{\$3,500,000} \times 100 = 11.9\%$$

Based on the guidelines followed by bonding firms, Eastway is operating with a very comfortable level of working capital. If it continues building a track record for strong financial performance, the firm will probably have no difficulty handling a larger volume relative to working capital in the future.

Leverage or Debt Ratios

The use of borrowed funds, or what is known as "leverage", allows profitable businesses to earn higher returns on owners' equity than would be possible without the use of creditors' capital. This premise assumes that the firm can earn a greater return with the borrowed funds than it must pay in interest. If the firm becomes unprofitable and earns a lower return than the interest rate, the losses on equity are quickly multiplied.

While the use of debt involves some risk to both the firm and its creditors, the rewards can be significant. In fact, most firms find they must use some debt in order to grow and remain profitable. However, the *level of debt* is of crucial importance in maintaining a position of financial strength. Debt ratios focus management's attention on the level of debt, repayment terms, and potential problems that may arise when using borrowed funds.

Each of the following ratios must be considered by a firm in light of its earnings, working capital, and cash flow. What may be a perfectly safe level of debt for one firm may be unreasonable for another due to differences in capital structure.

Earnings Coverage over Interest: This ratio reflects the number of times interest is covered by current earnings and the associated ability of the firm to service its current or additional interest payments. Creditors look at this coverage rate to ensure their loans are adequately protected and can be serviced easily from the firm's current earnings. The alternative reduction in equity that would be required if current earnings are insufficient is not a desirable condition for either the creditor or the firm.

In the case of Eastway Contractors, this ratio is derived as follows:

$$\frac{\text{Operating Profit}}{\text{Interest Expense}} \quad \frac{\$250,000}{\$\,8,000} = 31.3$$

The coverage of 31.3 times is far in excess of the upper quartile rate of 17.8 times, indicating the firm has minimized its borrowings in relationship to earnings. Creditors would be encouraged by these figures which indicate that the firm is not excessively leveraged and has plenty of capacity to incur additional debt.

Cash Flow Coverage Ratio over Current Portion Long Term Debt:
This ratio expresses the number of times the portion of long term debt that must be paid in the current period is covered by cash flow generated during the current period. Again, creditors want assurance that the firm's cash flow will be more than adequate to pay current period principal on outstanding loans.

$$\text{Cash Flow Coverage over Current Portion of Long Term Debt} = \frac{\text{Net Profit \& (Depreciation \& Amortization)}}{\text{Current Maturity of Long Term Debt}}$$

This ratio for Eastway is computed as follows, using the above formula and data from Figures 3.1 and 3.2.

$$\frac{\$260,000 \text{ \& } \$20,000}{\$15,000} = 18.7$$

The coverage rate of 18.7 times is well in excess of the 9.9 for the upper quartile, which is another indicator of the unusually low level of debt carried by Eastway. If a need arises, the firm would be well situated to borrow additional funds.

Net Worth to Fixed Assets Ratio: This ratio evaluates the portion of an organization's net worth committed to undepreciated fixed assets. A higher ratio represents a more significant investment in fixed assets. Investments in fixed assets cause greater fixed costs, such as depreciation and maintenance, which in turn causes a higher break-even point. If sales revenues are reduced while the costs of fixed assets remain constant, the firm will likely suffer losses and working capital shortages.

$$\frac{\text{Portion of Equity Invested in}}{\text{Fixed Assets}} = \frac{\text{Net Fixed Assets}}{\text{Net Worth}} \times 100$$

The portion of Eastway's equity invested in fixed assets is derived as follows:

$$\frac{\text{Net Fixed Assets}}{\text{Net Worth}} \quad \frac{\$85,000}{\$470,000} \times 100 = 18\%$$

Therefore, 18% of the firm's net worth is committed to undepreciated fixed assets. It is likely that new and less established firms will exhibit poorer performance in this ratio than the more established firms that have acquired and depreciated their assets over time while building greater net worth.

Debt to Worth: This relationship, often referred to as the Safety Ratio, compares the level of capital contributed by creditors to the equity of the firm. A higher ratio indicates a heavier dependency on debt, with the associated risk to creditors. A lower ratio reflects greater long term financial strength and the ability to assume greater liabilities in the future.

$$\text{Debt to Worth} = \frac{\text{Total Liabilities}}{\text{Net Worth}}$$

Eastway's *debt to worth ratio* is computed as:

$$\frac{\text{Total Liabilities}}{\text{Net Worth}} \quad \frac{\$515,000}{\$470,000} = 1.1$$

Eastway's liabilities are therefore 1.1 times greater than its net worth, which is very favorable compared to the median of 1.9, and equal to the upper quartile. While this is a safe position, the firm should continue efforts to limit its liabilities and increase its net worth.

Profitability Ratios

The primary objective of most business entities is to earn a profit. Without a reasonable level of profitability, an organization would be unable to grow and may ultimately be driven out of business. Management performance in earning profits is commonly evaluated through operating ratios. These ratios express the relationship of financial returns to either total revenues or the various resources available to the firm. The more efficient the firm, the higher the returns are likely to be. Comparison of a firm's profitability to the industry as a whole during any given period is a clear barometer of the firm's relative current financial performance as well as an indication of likely performance in the future.

Gross Profit: To realize net profits, a contractor must first earn revenues in excess of its *cost of sales* (labor, material, subcontractors, and equipment). This excess sum, or gross profit, is then allocated as necessary to indirect costs, general and administrative expenses and, finally, net profit. The gross profit ratio reflects the percentage difference between revenues and the associated cost of sales.

$$\text{Gross Profit Ratio} = \frac{\text{Gross Profit}}{\text{Revenue}}$$

In the case of Eastway, the gross profit percentage is determined as follows:

$$\frac{\text{Gross Profit}}{\text{Revenue}} \quad \frac{\$860,000}{\$6,000,000} \times 100 = 14.3\%$$

The Eastway gross profit margin is slightly lower than the average of 14.8%, indicating that the firm has been successful in estimating and completing its projects with a comfortable level of gross profit. In order to remain competitive, it is important that direct costs plus gross profit not exceed that of other contractors in the market. Above-average net profits will most likely result from carefully controlling the level of operating expenses and pursuing a level of sales most appropriate for the firm.

Net Profit: After direct costs are deducted from revenue, yielding gross profit, certain operating expenses must then be deducted from gross profit to yield *net profit*. These operating expenses include indirect field costs as well as general and administrative expenses. The net profit ratio is the difference between the revenue and all costs, divided by the revenue.

$$\text{Net Profit Ratio} = \frac{\text{Net Profit Before Taxes}}{\text{Revenue}} \times 100$$

For Eastway, net profit is computed as follows:

$$\frac{\text{Net Profit Before Taxes}}{\text{Revenue}} \quad \frac{\$260,000}{\$6,000,000} \times 100 = 4.3\%$$

The 4.3% net profit earned by Eastway is considerably higher than the median level of 2.9%, indicating an unusually strong return on sales.

Return on Equity: This ratio reflects the ability of management to earn a favorable return on the owners' investment, or equity, in the firm. Returns on such investments as government bonds and money market certificates are typically considered the break-even point of acceptability. The risk of operating a business should yield considerably higher returns on investment than conservative and low return instruments.

$$\text{Return on Equity} = \frac{\text{Profit Before Taxes}}{\text{Net Worth}} \times 100$$

Eastway's performance in this regard is derived as follows:

$$\frac{\text{Profit Before Taxes}}{\text{Net Worth}} \quad \frac{\$260,000}{\$470,000} \times 100 = 55.3\%$$

This is an extremely favorable return, due largely to the high return on a relatively large volume of sales as compared to the firm's net worth. As many firms become more established and better capitalized, net worth often increases relative to sales (and the associated potential profits), such that the return on equity stabilizes at a somewhat more conservative level.

While a high return on equity may indicate superior performance, it could also suggest that the firm is attempting too much volume with an uncomfortably low level of working capital. On the other hand, a low return may indicate either sub-par performance or a more conservative managerial approach, with an unusually high level of working capital. Therefore, as in all cases, this ratio must be evaluated in conjunction with other data in order to understand the true performance of the firm.

Return on Assets: This ratio expresses the effectiveness of the firm in earning a favorable return on all the resources available to it, both borrowed and invested. A higher ratio indicates success in employing the firm's assets to earn the best profits possible. Distortions may occur in this ratio if a firm has an uncommonly high level of depreciated assets, unusual levels of income, or uncommon expenses in any given period.

$$\text{Return on Assets} = \frac{\text{Profit Before Taxes}}{\text{Total Assets}} \times 100$$

The return on assets earned by Eastway is computed as follows:

$$\frac{\text{Profit Before Taxes}}{\text{Total Assets}} \quad \frac{\$260,000}{\$985,000} \times 100 = 26.4\%$$

This return again reflects the firm's strong performance, particularly when compared with the upper quartile return of 15.6%. The favorable comparison is due largely to the firm's low level of fixed assets and high level of net profit on sales.

Cost Ratios

Cost ratios are used in analyzing cost activity related to the firm's ongoing operations. Cost ratios provide management with a means to control and plan the operating costs that ultimately determine the firm's competitive position and profitability.

As previously discussed, an effective cost accumulation system is crucial in developing meaningful cost information. Such a system enables management to classify and collect information related to the firm's direct, indirect, and general and administrative expenses. This information is used in the budgeting process to determine markups on direct costs sufficient to cover the firm's operating expenses. It is also invaluable in comparative analysis, allowing management to evaluate its success in controlling costs in successive periods and in relationship to other similar firms.

Cost ratios are commonly developed to reflect the portion of total revenue committed to the major cost categories, i.e., direct costs, indirect costs, and general and administrative expenses. However, managers will receive valuable insight into the firm's cost structure by analyzing the activity of individual cost classifications that are of specific importance to the firm. For instance, warranty expense may be tracked as a percentage of revenue to reflect the firm's success in controlling quality. Equipment expenses, interest, insurance, and employee expenses are a few other examples of specific expenses that managers might analyze through cost ratios. The Income Statement Trend Analysis shown in Figure 3.1 reflects such expense percentages for Eastway.

Direct Cost Ratio: This ratio reflects the portion of revenue expended on the *direct costs* incurred in the construction of buildings and site improvements. It may also be beneficial for contractors to track the activity of the individual direct cost categories (labor, material, subcontractors, and equipment). In fact, the total annual costs incurred for each of these direct cost categories is important information in allocating the firm's indirect costs, which will be explained in Chapter 4.

$$\text{Direct Cost Ratio} = \frac{\text{Direct Cost}}{\text{Revenue}} \times 100$$

The direct cost (sometimes referred to as *cost of sales*) ratio for Eastway in 1987 is computed as follows:

$$\frac{\text{Direct Cost}}{\text{Revenue}} \quad \frac{\$5,140,000}{\$6,000,000} \times 100 = 85.7\%$$

Given an average direct cost of 85.2% for contractors with revenues between $1 million and $10 million, Eastway's direct cost expenditures seem quite normal.

It should be noted that the gross profit percentage is what remains after the direct cost percentage is deducted from total revenue. In Eastway's case, this would be:

$$\text{Revenue } 100\% - \text{Direct Cost } 85.7\% = \text{Gross Profit } 14.3\%$$

Because a larger portion of revenue is expended on direct cost than on any other cost category, it offers managers the greatest potential for savings. If highly effective purchasing and other field control measures are implemented, any savings will drop directly to the bottom line. While it may be quite possible to cut 1% from annual direct costs, major adjustments would probably be required to cut 1% from the firm's annual operating expenses.

Indirect Ratio: The indirect cost ratio is an expression of the portion of revenue expended on total indirect costs. Indirect costs are those incurred as the result of field operations, but not necessarily traceable to individual projects. Examples include vehicles and equipment used on multiple projects, field-related insurance, and certain field employee benefits.

$$\text{Indirect Cost Ratio} = \frac{\text{Indirect Costs}}{\text{Revenue}} \times 100$$

Eastway's indirect cost ratio is determined as follows:

$$\frac{\text{Indirect Costs}}{\text{Revenue}} \quad \frac{\$130,000}{\$6,000,000} \times 100 = 2.2\%$$

Comparative indirect cost data for contractors is not readily available, for it is most often grouped with direct costs. However, contractors should monitor the trends that are related to both their total indirect costs and individual indirect cost classifications.

General and Administrative Cost Ratio: This ratio reflects the portion of revenue committed to general and administrative (G & A) expenses. These are expenses incurred in operating the business, but not related to the field. Examples include office employees, occupancy expenses, and marketing.

$$\frac{\text{General and}}{\text{Administrative Cost Ratio}} = \frac{\text{Gen. \& Administrative Expense}}{\text{Revenue}} \times 100$$

The G & A ratio for Eastway is derived as follows:

$$\frac{\text{G \& A Expenses}}{\text{Revenue}} \quad \frac{\$480,000}{\$6,000,000} \times 100 = 8.0\%$$

The combination of 8.0% for G & A expenses and 2.2% for indirect costs yields total operating expenses of 10.2% for Eastway in 1987. This compares quite favorably with the average of 12.3%.

Using Trend Analysis

A financial trend is the upward or downward movement of certain financial data in successive periods. Any selected individual account or group of accounts from the income statement or balance sheet may be tracked over time. Ratios can also be used to relate various accounts to one another. The objective of trend analysis is to identify positive and negative developments that affect the firm's financial strength over a number of periods. Equipped with this information, management can effectively respond with any actions needed to correct potential problems and to perpetuate positive developments. In addition, trend analysis provides valuable information used in forecasting future financial performance, discussed in Chapter 4. Trends in income statement accounts, balance sheet accounts, and financial ratios are commonly tracked by managers, who may also track non-financial data, either independently or in combination with financial data. For instance, changes from year to year in the dollar value of projects completed by each supervisor may indicate the effectiveness of the individual supervisors.

To ensure realistic measurement of trends, the data from successive periods should be evaluated to make sure it can be fairly compared. For example, if the firm has made any changes in its accounting methods, the data would require adjustments before it can be interpreted.

Displaying trend data in graphic form makes the analysis easy to grasp. Plotting the firm's performance and industry averages reveals significant trends. Graphs of Eastway Contractor's trends over three years are shown in Figures 3.5, 3.6, and 3.7. They illustrate the firm's increased financial strength over the three years.

A contractor's financial trends may be attributed to a wide range of factors. Some influences are internally generated, while others are the result of external factors. Management can control internal factors, but it has little control over external factors. It is possible that a firm's managers may take every reasonable internal measure to ensure success and still fail because of unforeseen or uncontrollable external factors.

Internal Factors Causing Trends

Employees: Employee productivity can have a significant impact on a firm's performance. Many managers view their employees as simply another resource, paying little attention to such organizational behavior issues as leadership, motivation, and satisfaction. Although these issues might not rise to the surface immediately through trend analysis, they may ultimately explain declining performance in such areas as profitability, revenues, and quality control.

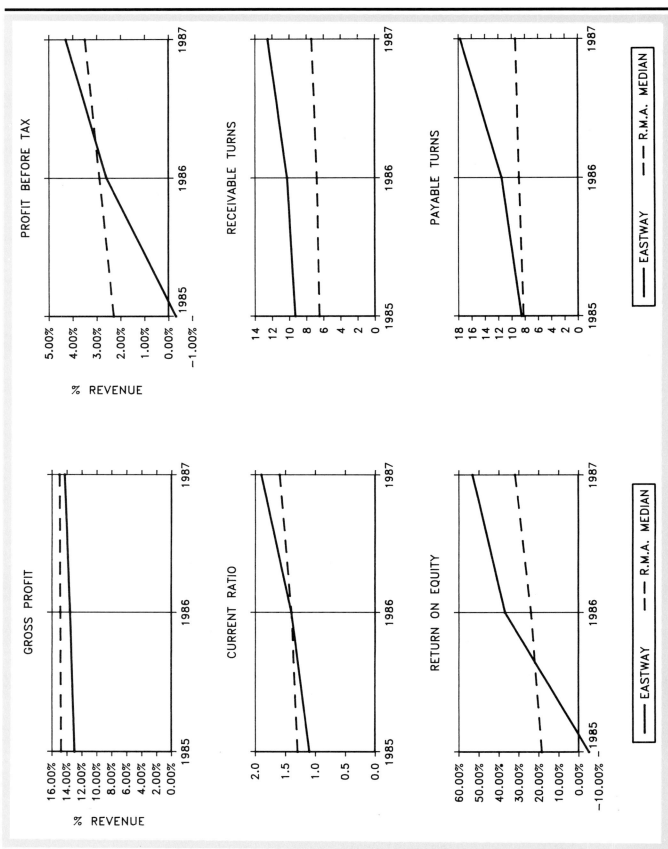

Figure 3.5

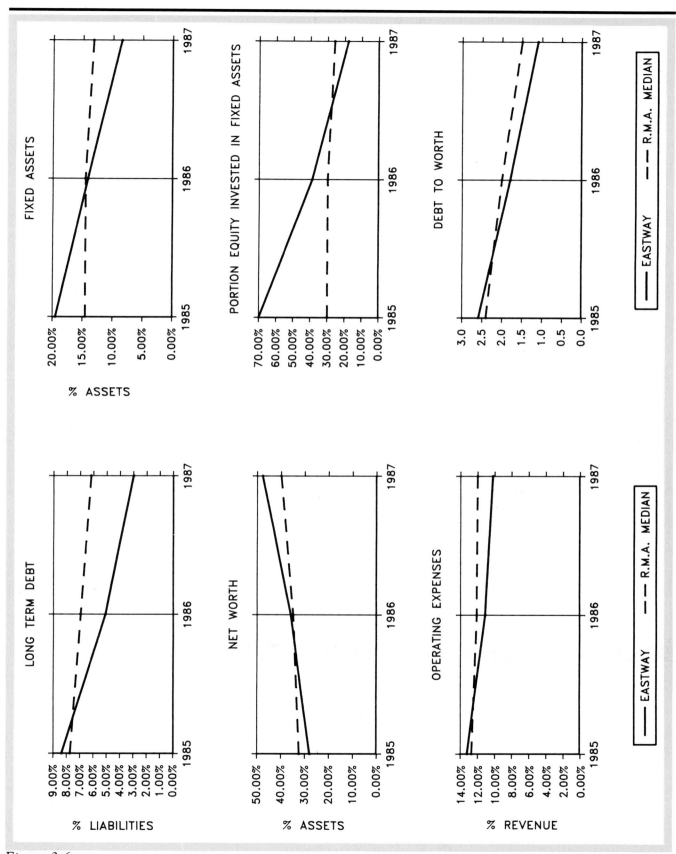

Figure 3.6

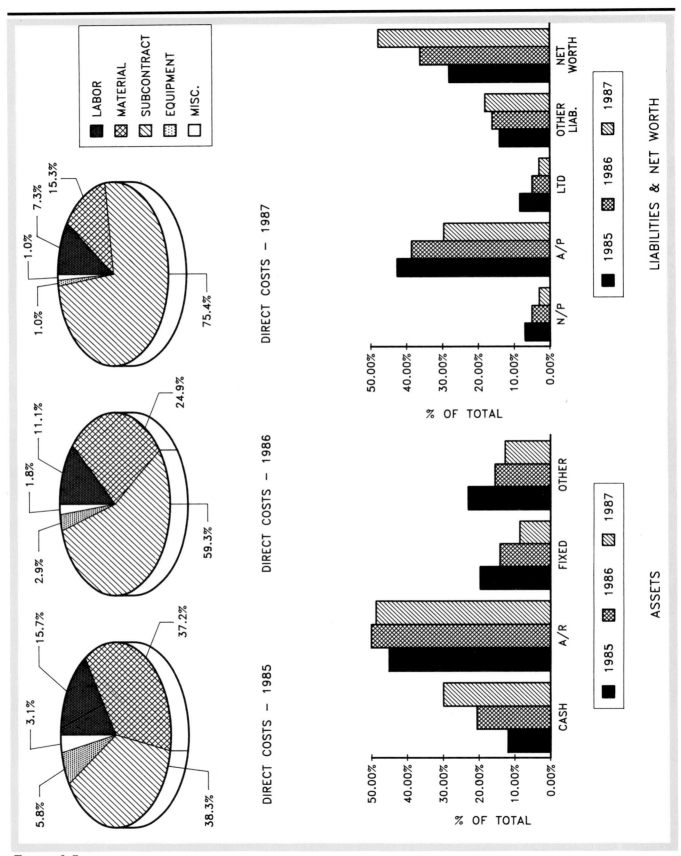

Figure 3.7

82

Marketing: Many contractors take a very unsophisticated approach to marketing their services. More aggressive and professional marketing should yield higher revenues at higher margins of profit.

Cost Control: As a firm moves from the Upstart stage to the Strike Force stage, it becomes adept at controlling its direct, indirect, and general and administrative expenses. Furthermore, as volume increases, the level of operating expenses relative to volume should decline, while the percentage of direct cost increases.

Receivables and Payables: In the Upstart stage, the firm may collect its receivables and pay its vendors later than desirable. However, most successful firms soon learn to control receivables and payables to their best advantage.

The ability to respond promptly to vendor invoices is another advantage of demanding prompt payment of receivables.

Debt Management: There is an evolutionary process that occurs in the debt structures of most contractors, particularly in the years of formation. Because of the high level of fixed assets often required in contracting, most firms incur a high level of debt when first starting the business. At the same time, the firm's net worth and working capital might be at minimal levels.

As time passes, and assuming reasonable profitability, most firms experience a reversal of these conditions. Long term debt is reduced as the firm pays off equipment, and net worth increases as the earnings are retained in the firm. While most firms must add new debt to acquire additional fixed assets, the relationship of debt to equity, earnings and cash flow is diminished. Many firms find that the portion of direct costs for materials and labor (as opposed to subcontractors) also diminishes. As the size of projects and total volume increase, contractors often depend more on subcontractors and need fewer fixed assets, such as shop space and equipment.

External Factors Causing Trends

Competition: As new construction firms enter the market and/or the level of building activity declines, building competition for new projects often becomes intense. Some firms make mistakes in their estimates or do not have a clear understanding of their true costs of doing business. As a result, the gross profit margins in the industry as a whole or any given region may also decline. Unfortunately, this condition is beyond the control of the individual contractor, who must continue operating even if his level of profitability is reduced.

Technological Improvement: New products and equipment have been introduced into the building industry which allow greater efficiency. One example is computer programs providing more efficient financial and project control. Contractors who take advantage of such advances will be more competitive and earn more consistent profits.

Economic Influences: The fluctuations in national and regional economic activity have a direct impact on the demand for building projects. Inflation, the stock market, interest rates, and the value of the dollar all play directly into this economic equation. In addition, as interest rates rise, the contractor's cost of borrowing is increased, causing reduced profits.

Demographic Influences: As the level and distribution of population in any given region or local market changes, there is a direct impact on the building industry. As population increases, additional housing is naturally required and commercial services soon follow. If an area becomes a business center, demand will grow for office and industrial space.

Government Regulations: Building construction is one of the most highly regulated industries in the U.S. As new environmental, building and employment requirements are implemented, contractors may experience shifts in their financial structure. Changes in taxation play a significant role in a contractor's performance. What may be an ideal financial structure under one set of tax rules may be quite different under another.

Trend Analysis —Eastway

The hypothetical building firm of Eastway Contractors, introduced briefly in Chapter 2 with its 1987 financial data, is used as an example in this Chapter's discussion of financial ratios. The data for the three-year period from 1985 through 1987 is used in the following example on trend analysis.

Eastway was incorporated in 1983 as a small residential building firm, but moved quickly into the light commercial market. By 1985 the firm had built its revenues up to the $3 million level, but was experiencing very poor financial performance. After addressing the causes of its financial ills in the subsequent two years, the firm was able to reverse its financial structure and earn superior profits.

The approach taken by Eastway in planning and implementing its strategies is discussed in the following sections. References are made to the computer-generated financial data in Figures 3.1, 3.2, and 3.3. The format used for Eastway's financial trend analysis is recommended for all contractors. Such statements should be updated yearly. A monthly analysis of the same information is also recommended, and should be easy to perform with the proper financial control systems.

Income Statement Trends

Eastway Contractors experienced 33% growth in revenue from 1985 to 1986, and 50% from 1986 to 1987. These increases were due to the strong regional economy in its market and a very aggressive marketing effort. The firm focused on pre-engineered steel buildings as its market niche, and had become well-known for competitive prices, timeliness, and high quality. As a result, the firm acquired the majority of its projects on a design/build negotiated basis, in contrast to the less profitable competitive bid basis.

Direct Costs: From 1985 to 1987, Eastway saw its direct costs go from 87.0% of revenue to 85.7%, with the associated gross profit increasing from 13.0% to 14.3%. At 85.7%, the firm's direct costs were very close to the 1987 median of 85.2%. The decrease in direct cost is indicative of the firm's efforts to improve its estimating and purchasing procedures. However, it should be noted that many well managed firms involved in large projects often must compete at considerably lower percentages of operating expenses and profits. This situation can be seen in Figure 3.4 under the column for contractors doing between $10 million and $50 million in sales. Fixed operating costs as a percentage of sales are generally reduced as the level of sales increases.

It is interesting to note that the distribution of total direct costs among the major direct cost categories (labor, subcontractors, material, equipment, and miscellaneous) changed dramatically from 1985 to 1987 for Eastway. Most importantly, the percentage of direct cost for labor and material dropped, while the percentage for subcontractors increased. In its early years, the firm had hired tradesmen and provided them with the necessary equipment to perform such trades as masonry, concrete, and carpentry. As the availability of skilled workers diminished and government regulations on employment increased, the firm found it far more advantageous to subcontract the majority of its trade work.

Indirect Costs: With the increase in volume and greater operational efficiency, the firm's indirect costs declined from 4.7% of revenue in 1985 to 2.2% in 1987. As the firm began using more subcontractors, the percentage of revenue committed to such indirect field costs as equipment and vehicles diminished.

General and Administrative Expenses (G & A): As in the case of indirect costs, the increase in volume and improved managerial controls has allowed the firm's general and administrative expenses to decline from 8.5% of revenue in 1985 to 8.0% in 1987.

The expenditure for office employees compared to revenues had been determined too large for 1985, reflecting a general ineffeciency in managing the administrative affairs of the business. With better control systems, including computerization, Eastway now has higher revenues and lower office employee expense.

The salary of $100,000 drawn by the owner/president of the firm in 1985 was excessive given that year's revenue of $3 million. The increase to $150,000 by 1987, however, can be considered reasonable for sales of $6 million.

Office equipment expenditures grew at a faster rate than revenue due to the investment in highly efficient computer equipment and additional office furniture.

The more aggressive marketing effort caused expenditures to increase from 0.1% of revenue in 1985 to 0.6% in 1987. The expenditure of just $4,000 in 1985 was far too low, reflecting management's "Upstart" vision that plenty of highly profitable work would somehow magically fall into their laps. The $35,000 marketing investment in 1987 is much more realistic for a firm seeking continued growth in revenue and profitability.

Expenditures for professional fees increased from $3,000 in 1985 to $12,000 in 1987, due primarily to management's recognition of the need for professional guidance from attorneys, financial advisors, and marketing specialists.

Office supply expenses as a percentage of revenue were somewhat out of control in 1985 at a level of $12,000, increasing only to $15,000 in 1987 despite the 100% increase in sales. During that time management became much more conscious of waste and purchasing control.

Profit Before Taxes: Although the gross profit of 14.3% in 1987 was still somewhat shy of the median of 14.8%, the net profit before taxes actually increased from 0.3% to 4.3% during this period. This rise in net pre-tax profit is due to the decrease in total operating expenses relative to sales volume.

Balance Sheet Trends

Current Assets: Eastway carried current assets of $495,000 at the end of 1985, or 69.2% of total assets of $715,000. By 1987, current assets of $860,000 represented 87.3% of total assets of $985,000. Given the industry average in 1987 of 77.7%, it can be seen that the firm was more committed to fixed assets than average in 1985, but less committed by 1987.

Most importantly, the portion of total assets attributed to Cash and Equivalents increased from 11.9% of total assets in 1985 to 29.9% in 1987, which is well above the average of 18.5%. This reflects management's commitment to good receivable control and minimal investment in fixed assets.

Accounts Receivable: Progress Billings stands at 40.6% of total assets in 1987, versus 35.9% in 1985. This is not a positive trend, and should be investigated, with corrective action taken. On the other hand, Current Retention Receivables has been reduced from 9.1% to 8.1%, proving management's success in negotiating more favorable retention terms with owners and lenders.

In 1985 the firm held $25,000 worth of inventory in miscellaneous materials in its shop. It soon learned that this investment was unwise due to carrying costs, damage, and obsolescence. By 1987 it had reduced this investment to $10,000, or just 1.0% of total assets, well below the average of 2.7%.

Costs & Earnings Exceeding Bills were reduced from 7.7% of total assets in 1985 to 5.1% in 1987, which is slightly less than the industry average of 5.5%. It is in the best interests of the firm to bill for all costs as soon as possible. Payment terms must be negotiated in advance and draw requests issued at regular intervals to ensure this occurs.

Non-Current Assets: The most significant movement within Non-Current Assets is the reduction in Fixed Assets. In 1985 Eastway was carrying $140,000 of undepreciated Fixed Assets, which represented 19.6% of Total Assets. Management invested heavily in large field equipment in its early years, and employed field crews to perform much of the tradework. The firm thereby incurred an undesirably high level of debt, causing an excessive level of interest payments.

By 1987, Eastway had sold much of the equipment, reducing its undepreciated Fixed Assets to $85,000, just 8.6% of total assets. This is a positive trend, which should be carefully monitored by the firm at all times. Asset management issues are discussed in Chapter 6.

In 1985, the firm's balance sheet reflected a $50,000 investment in Joint Ventures and Investments, which in this case represent real estate development partnerships. It is generally more advantageous to make such investments personally or through an independent development entity controlled by the contractor, rather than through a construction firm. Tax incentives are usually more favorable on a personal level. Personal investments have no impact on the firm's financial structure, and such assets are protected from potential lawsuits against the firm. For these reasons, the firm's financial advisors suggested that the owner either sell the real estate to other parties or acquire it personally.

Current Liabilities: The firm's increased liquidity and decreased debt has brought about changes in its liability and owner's equity accounts from 1985 to 1987. The result is a more resilient financial structure, less exposed to losses in the event of an economic downturn or excessive competition.

Notes Payable of $50,000 in 1985 represented 7% of total liabilities and equity. By 1987 this account had dropped to $30,000, or 3.0% of the total, despite a doubling in revenue. This percentage is less than half of the median figure of 6.2%, a strong indicator of liquidity. The increase in working capital results in a reduction in dependency on short term loans.

While the level of Accounts Payable Trade dropped only from $285,000 to $265,000 during the three year period, the percentage relative to Total Liabilities and Net Worth dropped from 39.9% to 26.9%. Given the median of 30.2%, the firm has been successful in reducing its payables to a very acceptable level.

The Current Maturity Long Term Debt account has dropped from $30,000 to $15,000, or 4.2% of the total vs. 1.5%. This is the most direct indicator of the firm's commitment to reduced debt, both current and non-current.

Non-Current Liabilities: The Long Term Debt account has also been reduced by 50%, from $60,000 to $30,000. When combined with the Current Maturity LTD account, it can be seen that total debt dropped from $90,000 to $45,000 from 1985 to 1987. On a percentage basis, total debt dropped from 12.6% of total liabilities and net worth to just 4.5%. Since the median percentage for total debt is 10.0%, the firm has clearly performed exceptionally well in this regard.

Net Worth: As a result of the downward trends in most liability accounts, net worth has increased from 28.0% to 47.7% of the total. Given the median of 35.4%, the firm is better capitalized than most contractors.

Ratio Trends

As discussed in prior sections, Eastway recognized several flawed features in its financial structure and performance in 1985, and got on a course to correct them by 1987. These efforts improved the trends in ratios from 1985 to 1987. In practically all cases, the individual ratios of Eastway are above the median in 1987, and better than the upper quartile in most cases.

Financial Cause and Effect: In evaluating financial strength, managers and external parties typically attach the most significance to the level of *working capital* and *profitability*. A strong showing in these two areas, along with good expectations of future revenues, will go far in establishing and maintaining the confidence of external creditors. On the other hand, a shortfall in either of these areas will immediately send up red flags, which in turn has a negative effect on creditor relations. After a trend and ratio analysis of the firm's financial performance, managers must investigate the causes of any negative results.

A wide range of factors influence the level of a firm's profitability and working capital, as shown in Figure 3.8. While a number of other less obvious factors can certainly have an impact, this cause and effect diagram illustrates the most common factors causing negative trends in financial performance.

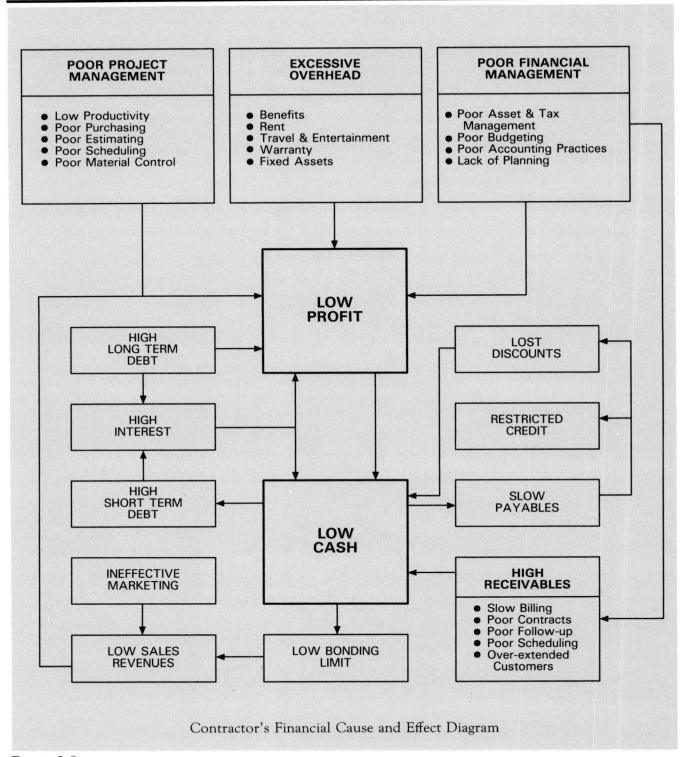

Contractor's Financial Cause and Effect Diagram

Figure 3.8

Eastway's financial shortfalls in 1985 can easily be recognized in this diagram. A short summary of the most significant issues follows.

1. The firm was operating with far too little working capital relative to revenue, and experiencing cash flow problems. Both receivable and payable turnover was poor, causing further cash flow problems.
2. The firm had become much too dependent on debt, both short and long term. This was due in part to excessive investment in fixed assets with the associated long term debt and interest expense. The interest, in turn, drained working capital and profits.
3. Poor estimating and purchasing controls caused a low gross profit, and excessive operating expenses resulted in a loss on the bottom line. This loss caused a negative return on equity and assets.

Chapter Four

FINANCIAL FORECASTING

Chapter Four

FINANCIAL FORECASTING

Accounting and financial analysis focus on the results of past business decisions pertaining to operations, financing, and investments. In contrast, financial forecasting deals with the projection of operating and financial performance into *future* periods. Forecasting is a key element in business planning and control. In this process, the contractor positions the firm's future activities based on anticipated external conditions. Such conditions include the regional economy, competition, and local demand for commercial construction—major factors that directly affect the firm's performance and must be reflected in its plans and objectives.

Unfortunately, the critical impact of accurate financial forecasting on the long term success of the firm is sometimes overlooked. Drawing on their construction-oriented background, many contractors tend to concentrate on project management functions while neglecting financial control functions. Without proper attention, future periods hold the potential for cash flow problems, insufficient profit margins, or a host of other financial pitfalls. On the other hand, careful projection of both marketplace conditions and the likely performance of the firm given those conditions allows the prudent contractor to avoid many difficulties.

The Planning and Control Process

Two of the management functions most crucial to the success of any firm are *planning* and *controlling*. Planning is required at all levels of the organization involving decisions as to what should be done and how it is to be done. The supervisor scheduling material deliveries and subcontractors for the following week is engaged in planning. The president is also planning when establishing new corporate goals or policies.

Control is the follow-up effort required to ensure that operational results conform to the plan. The collection and evaluation of accounting data is an example, in this case, a *financial control* function. The enforcement of proper construction techniques is a *quality control* function.

Three types of planning and control processes are found in most successful contracting organizations:

1. Strategic Planning
2. Managerial Control
3. Task Control

Figure 4.1 shows the wide variety of specific activities involved in each of these processes. It is helpful to understand the position of financial forecasting within the overall planning and control process.

Strategic Planning

At the strategic planning level, top management establishes the future goals of the firm and specific strategies to accomplish those goals. Relevant issues include projected sales volume, type and location of building projects, improvements to organizational structure, refinements to managerial systems, and specific financial goals. The contractor's business environment is never static. As a result, goals and strategies will change continuously in response to new internal and external influences. A business plan for our example contracting firm, Eastway Contractors, is shown in Appendix B. This plan covers the goals, strategies, and policies that the firm has determined to be most crucial to its success in 1988. Also included in Appendix B is a suggested format for the development of a business plan.

Managerial Control

Managerial control is the process whereby management compels both individuals and the organization as a whole to implement the firm's strategies in an effective and efficient manner. This process involves the development of formal and informal procedures, communicating specific responsibilities and performance standards, follow-up observation, and the interpretation of various reports. Two major aspects of managerial control are the financial reporting and forecasting functions, which allow management to monitor current performance and predict future performance. In a sense, this is where "the rubber hits the road", for clear, "bottom line" results are the output of these functions.

Task Control

The third planning and control process, task control, involves ensuring that specific tasks are carried out in accordance with designated procedures. Task control is particularly important in project management functions. Many daily decisions are made in this area that have an impact on the financial performance and liability of the firm. Examples include scheduling subcontractors, purchasing materials, and ensuring quality control. However, such administrative tasks as bookkeeping, preparing contracts, and consistent work flow must also be carefully controlled.

Responding to Projections

As various reports are generated at the managerial control level, management must continually evaluate the results and respond with new projections, strategies, and corrective actions. For instance, an unexpected drop in sales volume may require either a new marketing strategy or simple acceptance of reduced volume. If reduced volume seems inevitable, the firm's organizational and financial structure must be adjusted accordingly to avoid excessive operating expenses relative to sales volume. This planning, evaluation, and correction process is the essence of business management, requiring constant effort on the part of all managers within the firm.

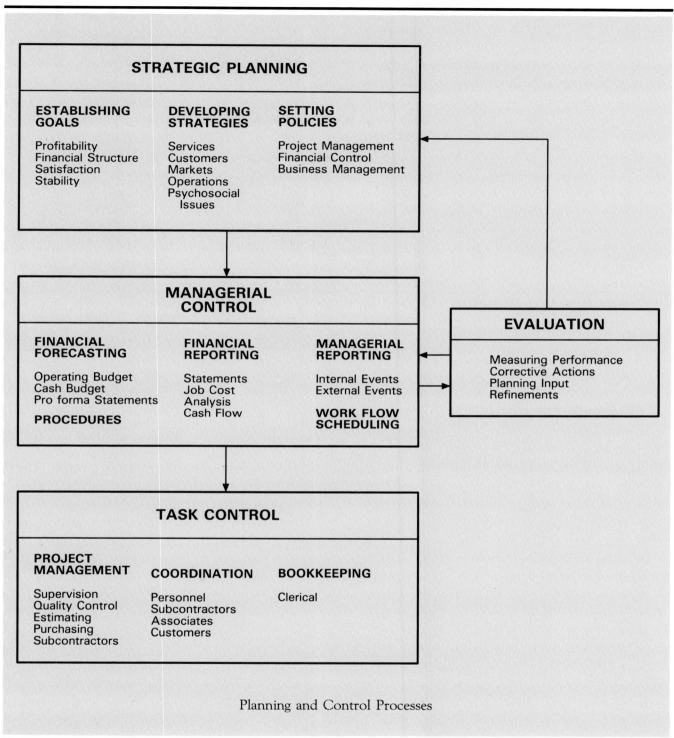

Planning and Control Processes

Figure 4.1

Financial Forecasting Methods

In forecasting financial performances, a variety of current and predicted financial data are input into a consolidated projection of future financial conditions. These anticipated conditions are typically expressed in the form of the firm's operating budget, cash budget, and pro forma financial statements. The relationships of these inputs and outputs are illustrated in Figure 4.2. These forecasting techniques are clearly closely related; the assumptions underlying each must therefore be identical in order to arrive at a true and comprehensive picture of the firm's future financial condition.

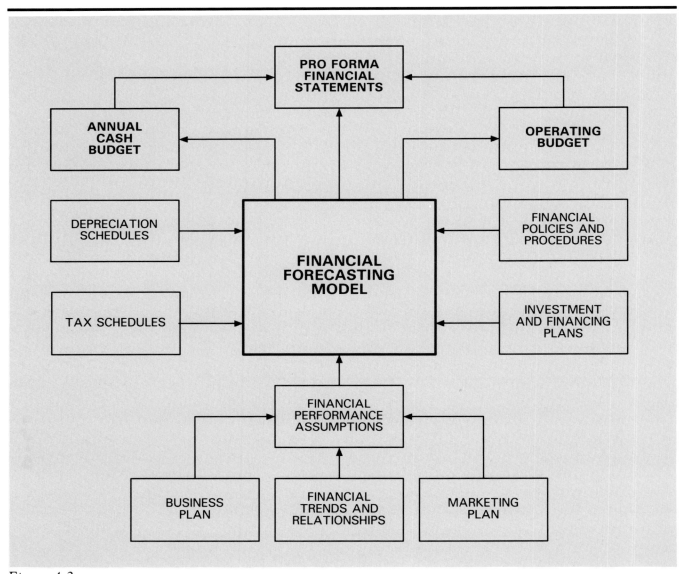

Figure 4.2

It is up to management to analyze the forecasted reports and consider modifications at the planning and/or control level that will bring about desirable improvements. For example, a consistent cash shortfall revealed by the cash budget may prompt management to renegotiate its line of credit, become more aggressive in its receivables policy, or tighten project schedules. Undesirable conditions revealed by the pro forma statements, such as the level of debt and associated interest expense relative to industry standards, may cause management to reconsider its investment in the fixed assets that drive up long term debt and interest.

While the budgets and pro forma statements described in this chapter can be produced manually, most contractors will find a computerized financial forecasting model far more expeditious. Such models are capable of quickly analyzing the effects of different levels of revenue, gross margins, operating expenses, investments, taxes, depreciation, and many other variables. While complex, fully capable software packages are available, simplified models can be developed in-house using a basic electronic spreadsheet.

Forecasting Input

Financial Performance Assumptions
The marketing plan provides important data for the financial forecasting model. General assumptions as to markets, sales volume, and profitability provide base information to which a quantity of more specific data is added. Current and recent financial reports also provide crucial data for the forecasting model. The trends found in various balance sheet and income statement accounts can be used to project likely performance and results in the future. Past performance as revealed through key ratios may also be used to project future trends.

Depreciation Schedules
As the firm acquires or disposes of fixed assets, or as depreciation schedules are modified, appropriate adjustments must be made to the model. A failure to accommodate true depreciation expenses will cause historical earnings figures to be overstated. The result may be insufficient gross profit margins which, over time, fail to produce enough cash to replace (at inflated costs) the assets consumed in the operations of the firm. Many firms, therefore, find that when older equipment becomes obsolete, they cannot afford to replace it.

Tax Schedules
Current tax schedules are crucial if accurate future net profits are to be reflected. The taxation of contracting firms tends to be extremely complex, as discussed in Chapter 8 on Tax Management. Consequently, much of this input must come from tax professionals.

Investment and Financing Plans
As the sales volume of a construction firm grows, a need evolves for some level of fixed asset investment. Improvements to office space, additional field equipment, or more sophisticated computer systems are examples of fixed assets that are normally expanded to accommodate higher sales volumes. The depreciation and maintenance expense associated with these fixed assets must be considered in any projection involving significant increases in sales. The method of purchase for such assets is also a major factor. If payment is made out of working capital rather than in the form of new, long term debt (with the associated interest expense), there is a significant impact on all three forms of forecasting reports.

Financial Policies

The accounting and financial management policies practiced by the firm play a direct part in the firm's performance over time. The choice of financial and tax reporting methods will certainly affect future reported performance. The treatment of receivables, payables, and debt, for example, has a direct impact on cash flow. Management's position on retained earnings and executive compensation will affect the pro forma balance sheet by either increasing or decreasing net worth.

Forecasting Output

Operating Budget

The most common financial forecast is the operating budget, an expression of the assumed operations during the forthcoming period (usually one year). This budget is essentially a projected income statement, including the sales volume, gross margin, operating expenses, and net profit anticipated during the period. The development of an operating budget is the major focus of this chapter.

Cash Budget

The cash budget is a detailed projection of cash flow, typically on a monthly basis. The focus is on the incidence of cash receipts and disbursements, and the potential need for short term borrowing at any time during the course of the projected period. The specific procedures involved in developing a cash budget, as well as a number of other factors influencing cash flow are discussed in Chapter 5, "Cash Management".

Pro Forma Statement

The most comprehensive method of understanding future conditions is by developing pro forma statements. These are standard income statement and balance sheets extended into the future with all anticipated adjustments resulting from operations during the forthcoming period. The operating and cash budgets, as well as planned capital expenditures, provide key data for developing forecasted financial statements. Pro forma statements for Eastway Contractors' expected 1988 financial results are included in Appendix B.

Benefits of Financial Forecasting

These three forecasted reports play a number of critical roles in directing and monitoring the activities of the firm, including:

1. **Planning.** Developing budgets helps to refine the decisions made at the strategic planning stage. Potential conflicts between marketing and field production capability may be highlighted, allowing management to build or reduce the field staff in a smooth manner. Possible cash shortfalls can be anticipated and planned for well in advance, when the firm is in the best position to negotiate with the bank.

2. **Communication.** Financial forecasts provide a valuable tool for use in communicating specific goals and responsibilities to top management. Capital expenditures, sales goals, cash flow expectations, and other crucial assumptions must be related to the responsible managers.

3. **Control.** Budgets provide a statement of desired results, against which actual performance can be measured. When shortfalls occur, budgets direct attention to the areas where action may be needed. Evaluation of the success of the firm and/or individual managers through comparison with budgets is largely an objective effort. Without such "hard" guidelines, managers are left to evaluate performance subjectively, on the basis of perceptions.

4. **Education and Motivation.** It is important to the budgeting process that managers are informed of the role that they and their respective "responsibility centers" play in the firm's overall operations. An awareness of budget assumptions tends to motivate managers to work toward meeting their individual guidelines.

Cost Concepts

In analyzing and predicting the expenses of a construction firm, one must understand basic cost concepts. In fact, it is a lack of this kind of understanding that leads to many causes of construction failure. Many contractors do not have a grasp of their true costs of doing business. Furthermore, various cost components change at different rates in response to increases or decreases in sales volume. As a result, many managers do not really know what mark-ups are required to earn a profit, nor do they know when they are making or losing money. Finally, the lack of appropriate job cost systems prevents many firms from knowing which projects are resulting in profits or losses.

Cost is a monetary measurement of the amount and value of various resources used for some purpose. In the case of construction, we have the following major cost categories:

Direct Costs
Direct Costs are incurred in the actual field production of building structures and site improvements. They include the labor, material, subcontractor, equipment, and miscellaneous expenses associated with specific projects. These expenses are directly incorporated into the physical improvements. An estimate form outlining the major divisions of direct cost in light commercial projects is provided in Appendix B.

Indirect Costs
Indirect costs are incurred as the result of field operations, but cannot be clearly or easily attributed to specific projects. They are not part of the physical improvements. Examples include certain field employee benefits, field equipment that is used on several projects, and vehicles driven by supervisors who are responsible for more than one project. These costs are allocated, or spread, among all projects.

General and Administrative Costs
This cost category is often referred to as *overhead*. These costs are not field-related. They are instead associated with operating the business, and cover such items as office rent, administrative salaries, marketing expenses, and insurance. General and administrative costs are allocated to all projects on the basis of the relationship of each project's dollar value to total annual sales.

Analysis
The total price that contractors bid for any given project is most often derived by applying certain mark-ups to estimated direct costs. The mark-ups must be of a sufficient level to recapture the indirect, as well as general and administrative expenses associated with each project, while

also earning a reasonable profit. Determining appropriate mark-ups is a major goal in preparing the operating budget.

The various indirect and general and administrative costs incurred by the firm can be classified as *fixed*, *variable*, or *semi-variable* cost elements. Each of these cost elements behave differently as the sales volume changes. Understanding and projecting all of the costs incurred by the firm over any range of anticipated sales volumes is the essence of effective budgeting.

Before defining each of these three cost elements in the following sections, it is important to draw an important distinction. In manufacturing or retail operations, any discussion of costs and income is generally reduced to the relationship of total costs (including direct costs) per manufactured unit, versus the income generated by each unit. However, contractors do not deal with such standardized units. Therefore, the following sections all relate to the operating expenses incurred by the firm compared to the total gross profit (generated at various gross margins and sales levels) that is available to cover operating expenses.

The cost data used in the following sections is drawn from Figure 4.3, a summary of cost and profit data for Eastway Contractors. The data covers the anticipated results of sales in the range of $4 million to $7 million, at gross margins from 13% to 16%. An electronic spreadsheet loaded with detailed cost information is used to project this data, which in turn guides the firm in establishing appropriate mark-ups. Developing this data is discussed in more detail later in this chapter.

Fixed Costs

Fixed costs do not change as the level of sales changes. Depreciation of field or office equipment is considered a fixed cost, for it does not change unless such equipment is sold or new equipment is purchased. Office rent and utilities are also fixed, for they remain constant regardless of sales volume. Most forms of business insurance are also treated as fixed costs.

Figure 4.4 is a cost-to-volume graph used to illustrate various types of cost behavior. In such a diagram, sales volume is customarily plotted along the bottom (x-axis), while cost and profit data are plotted along the vertical side (y-axis). A line on the diagram reflects the anticipated costs or profit at any level of volume. Fixed costs appear as a horizontal line, for they do not change as the sales volume changes.

Figure 4.4 suggests that total fixed costs for Eastway Contractors remain unchanged at $106,500 from $0 to $10 million in sales volume. This is not a truly accurate representation, for when sales volume drops below a certain level, management decisions will lead to a drop in the level of fixed costs. Management may lease smaller or less expensive office space or sell equipment in order to significantly reduce fixed costs. Likewise, as volume grows, certain "lumpy" cost increases may be required. For instance, at $10 million in sales, the president may decide that the firm can justify the purchase of a new backhoe and trailer. This new equipment will cause a significant increase in fixed costs.

As a result of either intentional decreases in fixed costs when sales plunge, or "lumpy" increases due to higher levels of sales, fixed costs are, in fact, only fixed within a certain range of sales volumes. This range is referred to as the *relevant range*, that range of sales volumes relevant to the specific circumstances facing the firm. Figure 4.5 illustrates this concept, showing a "stepping" of fixed costs at various sales volumes. In this example, the relevant range for budgeting purposes is from $4 million to $7 million in annual sales.

```
                            EASTWAY CONTRACTORS, INC.
                            NET INCOME FORECAST
                            RELEVANT RANGE $4M - $7M

--------------------------------------------------------------------------------
    FINANCIAL      - - - - - - - - - - - - - - GROSS PROFIT % - - - - - - - - - - - - - - - - - -
      DATA           13%                 14%                 15%                 16%
--------------------------------------------------------------------------------

          REVENUE  $4,000,000  100.00%  $4,000,000  100.00%  $4,000,000  100.00%  $4,000,000  100.00%
      DIRECT COST  (3,480,000)  -87.00%  (3,440,000)  -86.00%  (3,400,000)  -85.00%  (3,360,000)  -84.00%
                   ----------  --------  ----------  --------  ----------  --------  ----------  --------
     GROSS PROFIT     520,000   13.00%     560,000   14.00%     600,000   15.00%     640,000   16.00%
                   ----------  --------  ----------  --------  ----------  --------  ----------  --------
      FIXED COSTS    (106,500)   -2.66%    (106,500)   -2.66%    (106,500)   -2.66%    (106,500)   -2.66%
FIXED PORT. S. V. COSTS (212,000) -5.30%  (212,000)   -5.30%    (212,000)   -5.30%    (212,000)   -5.30%
VAR. PORT. S. V. COSTS  (118,200) -2.96%  (118,200)   -2.96%    (118,200)   -2.96%    (118,200)   -2.96%
   VARIABLE COSTS    (92,500)   -2.31%     (92,500)   -2.31%     (92,500)   -2.31%     (92,500)   -2.31%
                   ----------  --------  ----------  --------  ----------  --------  ----------  --------
NET PROFIT BEFORE TAX ($9,200)  -0.23%     $30,800    0.77%     $70,800    1.77%    $110,800    2.77%
                   ==========  ========  ==========  ========  ==========  ========  ==========  ========

          REVENUE  $5,000,000  100.00%  $5,000,000  100.00%  $5,000,000  100.00%  $5,000,000  100.00%
      DIRECT COST  (4,350,000)  -87.00%  (4,300,000)  -86.00%  (4,250,000)  -85.00%  (4,200,000)  -84.00%
                   ----------  --------  ----------  --------  ----------  --------  ----------  --------
     GROSS PROFIT     650,000   13.00%     700,000   14.00%     750,000   15.00%     800,000   16.00%
                   ----------  --------  ----------  --------  ----------  --------  ----------  --------
      FIXED COSTS    (106,500)   -2.13%    (106,500)   -2.13%    (106,500)   -2.13%    (106,500)   -2.13%
FIXED PORT. S. V. COSTS (212,000) -4.24%  (212,000)   -4.24%    (212,000)   -4.24%    (212,000)   -4.24%
VAR. PORT. S. V. COSTS  (148,550) -2.97%  (148,550)   -2.97%    (148,550)   -2.97%    (148,550)   -2.97%
   VARIABLE COSTS   (115,625)   -2.31%    (115,625)   -2.31%    (115,625)   -2.31%    (115,625)   -2.31%
                   ----------  --------  ----------  --------  ----------  --------  ----------  --------
NET PROFIT BEFORE TAX $67,325   1.35%    $117,325    2.35%    $167,325    3.35%    $217,325    4.35%
                   ==========  ========  ==========  ========  ==========  ========  ==========  ========

          REVENUE  $6,000,000  100.00%  $6,000,000  100.00%  $6,000,000  100.00%  $6,000,000  100.00%
      DIRECT COST  (5,220,000)  -87.00%  (5,160,000)  -86.00%  (5,100,000)  -85.00%  (5,040,000)  -84.00%
                   ----------  --------  ----------  --------  ----------  --------  ----------  --------
     GROSS PROFIT     780,000   13.00%     840,000   14.00%     900,000   15.00%     960,000   16.00%
                   ----------  --------  ----------  --------  ----------  --------  ----------  --------
      FIXED COSTS    (106,500)   -1.78%    (106,500)   -1.78%    (106,500)   -1.78%    (106,500)   -1.78%
FIXED PORT. S. V. COSTS (212,000) -3.53%  (212,000)   -3.53%    (212,000)   -3.53%    (212,000)   -3.53%
VAR. PORT. S. V. COSTS  (178,900) -2.98%  (178,900)   -2.98%    (178,900)   -2.98%    (178,900)   -2.98%
   VARIABLE COSTS   (138,750)   -2.31%    (138,750)   -2.31%    (138,750)   -2.31%    (138,750)   -2.31%
                   ----------  --------  ----------  --------  ----------  --------  ----------  --------
NET PROFIT BEFORE TAX $143,850  2.40%    $203,850    3.40%    $263,850    4.40%    $323,850    5.40%
                   ==========  ========  ==========  ========  ==========  ========  ==========  ========

          REVENUE  $7,000,000  100.00%  $7,000,000  100.00%  $7,000,000  100.00%  $7,000,000  100.00%
      DIRECT COST  (6,090,000)  -87.00%  (6,020,000)  -86.00%  (5,950,000)  -85.00%  (5,880,000)  -84.00%
                   ----------  --------  ----------  --------  ----------  --------  ----------  --------
     GROSS PROFIT     910,000   13.00%     980,000   14.00%   1,050,000   15.00%   1,120,000   16.00%
                   ----------  --------  ----------  --------  ----------  --------  ----------  --------
      FIXED COSTS    (106,500)   -1.52%    (106,500)   -1.52%    (106,500)   -1.52%    (106,500)   -1.52%
FIXED PORT. S. V. COSTS (212,000) -3.03%  (212,000)   -3.03%    (212,000)   -3.03%    (212,000)   -3.03%
VAR. PORT. S. V. COSTS  (209,250) -2.99%  (209,250)   -2.99%    (209,250)   -2.99%    (209,250)   -2.99%
   VARIABLE COSTS   (161,875)   -2.31%    (161,875)   -2.31%    (161,875)   -2.31%    (161,875)   -2.31%
                   ----------  --------  ----------  --------  ----------  --------  ----------  --------
NET PROFIT BEFORE TAX $220,375  3.15%    $290,375    4.15%    $360,375    5.15%    $430,375    6.15%
                   ==========  ========  ==========  ========  ==========  ========  ==========  ========
```

Figure 4.3

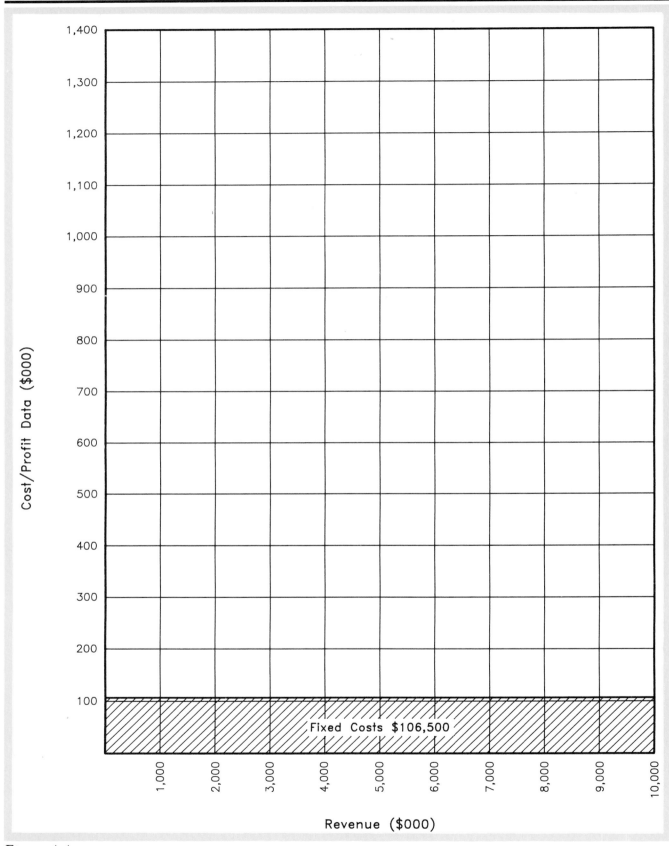

Figure 4.4

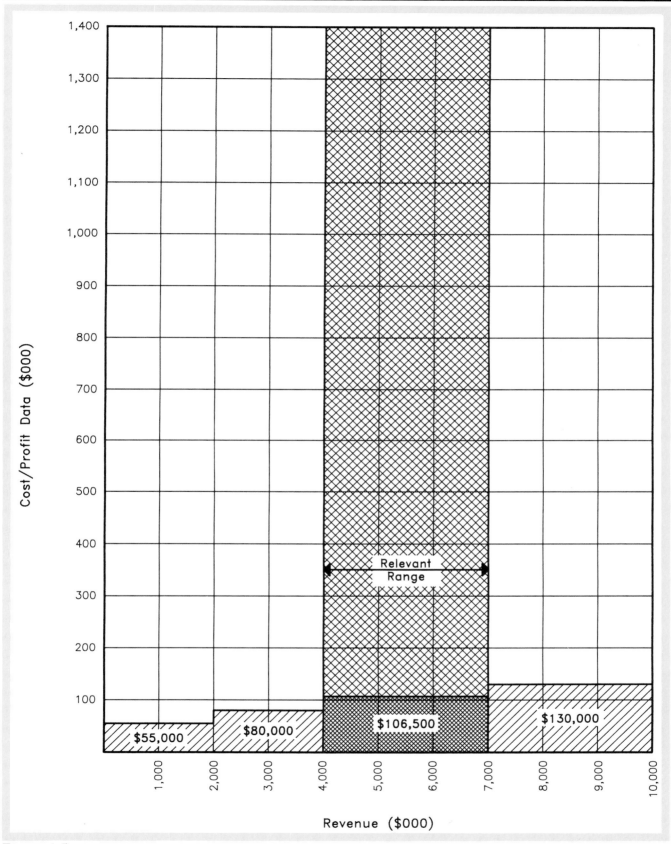

Figure 4.5

Semi-variable Costs

A cost which has both fixed and variable components is called a *semi-variable cost*. These costs change in the same direction as, but less than proportionately with, changes in sales volume. For example, a minimum number of employees must be retained regardless of the level of sales if the firm is to remain operational. The salaries and benefits associated with this minimum level would be fixed, while salaries and benefits for the additional staff required at a higher volume would be considered variable. Vehicles have fixed expenses in the form of depreciation and insurance, with variable expenses in fuel, oil, and maintenance. Equipment maintenance, telephone expense, and professional fees are other examples of costs with both fixed and variable portions.

Figure 4.6 illustrates the semi-variable cost concept. In this example, the fixed portion remains at $212,000 throughout all sales volumes, while the variable portion increases directly with sales volume. At $4 million in sales, the variable portion is $118,200, while at $7 million, the variable portion increases to $209,250.

Variable Costs

Costs that change directly and proportionately with changes in volume are referred to as *variable costs*. Examples include voluntary bonuses paid to employees, fuel and oil for field vehicles, and real estate commissions. If the firm had no sales volume, these costs would theoretically drop to zero. As volume grows, variable costs increase proportionately. For instance, if fuel and oil for field vehicles costs $4,000 at a sales volume of $4 million, and sales increase 50% to $6 million, fuel and oil will also increase 50%, to $6,000.

Figure 4.7 is a cost volume diagram that illustrates this variable cost relationship. At $0 in sales, there are no variable costs. However, in this example, at $4 million in sales, variable costs are $92,500. At $7 million in sales, a 75% increase, variable costs go up to $161,875, also a 75% increase.

Total Costs

The sum of fixed, variable, and semi-variable costs reflects the firm's total cost of operations. Figure 4.7 illustrates the combination of all such costs. As shown, at a sales volume of $7 million, the total cost can be broken down as follows:

Fixed Costs	$106,500
Semi-variable Costs	421,250
Variable Costs	161,875
Total Operating Cost	**$689,625**

Total variable costs at any given level of sales may be mathematically computed by first determining the portion of each dollar of gross profit expended for variable operating costs. This determination is calculated as follows:

$$\text{TVC/D} = \frac{\text{VC} + \text{VPSVC}}{\text{GP}}$$

where TVC/D = Total variable costs per dollar of gross profit
 VC = Variable costs
 VPSVC = Variable portion semi-variable costs
 GP = Gross profit

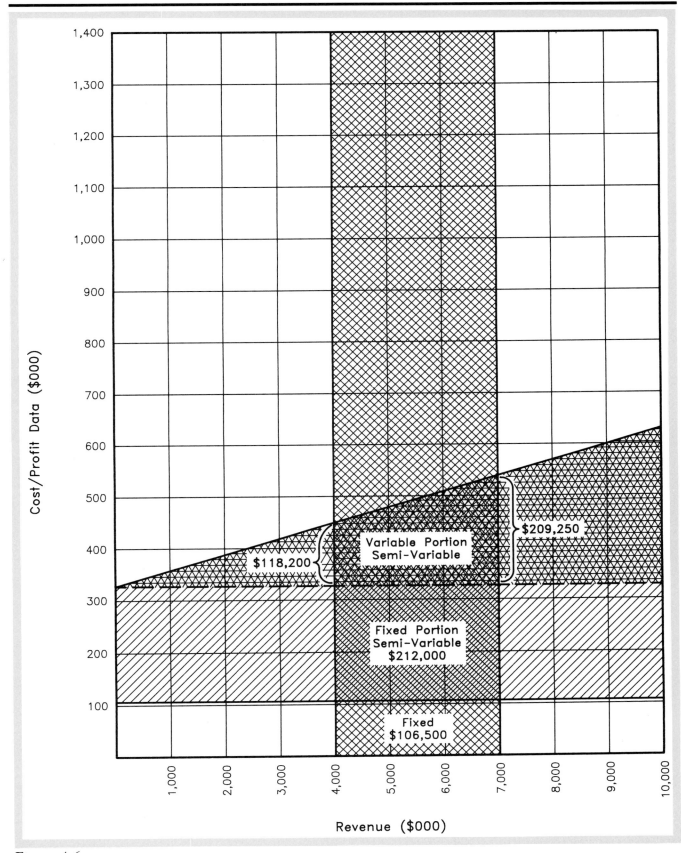

Figure 4.6

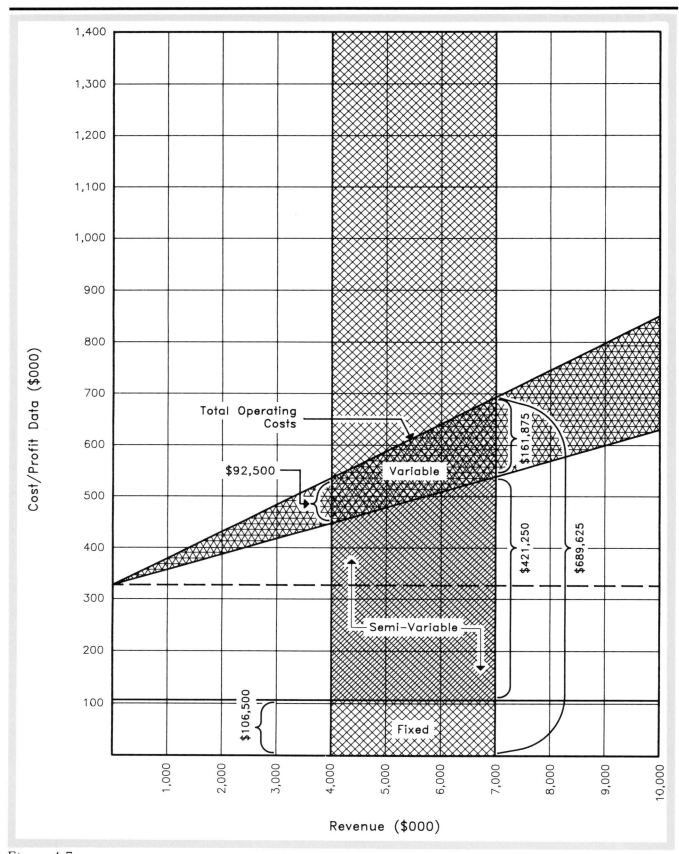

Figure 4.7

At the $7 million sales level, the following calculation would be used:

$$\text{TVC/D} = \frac{\$161{,}875 + \$209{,}250}{\$1{,}050{,}000} = \$0.3534$$

Therefore, approximately 35% of each dollar of gross profit earned is required to offset the variable operating costs of running the business. The remaining 65% is referred to as the *contribution margin*, which will be discussed in a subsequent section.

Given the value for total variable cost per dollar (TVC/D) of $0.3534, total costs can be computed as follows:

TC = FC + FPSVC + (TVC/D × GP)

where: TC = Total cost

FC = Total fixed cost (fixed costs plus fixed portion of semi-variable)

FPSVC = Fixed portion of semi-variable cost

In our example, this would be:

TC = $106,500 + $212,000 + (.3534 × $1,050,000)
= $318,500 + $371,070
= $689,570

The total cost derived, $689,570, closely matches (due to rounding) the total cost of $689,625 shown in Figure 4.7.

Break-Even Analysis

One final element to complete our diagram series is a slope line reflecting the gross profit earned at various levels of volume. The resulting diagram, shown in Figure 4.8, is often referred to as a *profitgraph*, for it graphs the profits or losses expected at various volumes. Given $0 gross profit at $0 sales, and a 15% gross margin at any sales level thereafter, the gross profit slope line shown is easily derived. At the point the gross profit line intersects the total cost line, the firm is exactly covering its total operating costs with earnings. In the example, this intersection occurs at a sales volume of approximately $3,270,000 and a gross profit of $490,000. A sales volume that is any lower will result in a loss; any greater sales volume will result in a profit. At the $7 million level previously considered, gross profit equals $1,050,000, while total cost equals $689,625, leaving a net profit of $360,375.

It should be noted here that management has a number of alternatives for altering the break-even point and the related cost-volume relationships. The most prominent choices are the following:

1. Increase sales volume by becoming more aggressive with the marketing effort.
2. Attempt to increase gross margin by charging a higher price, reducing operating expenses, or reducing direct costs. While a 1% savings in total costs will drop straight to the bottom line, it is far easier to realize that 1% through direct cost savings (85% of volume, on average) rather than through operating expenses (12% of volume, on average).
3. Reduce fixed costs by moving to smaller or less expensive offices, reducing the amount of equipment, or reducing the size of the staff.
4. Reduce variable costs by such means as reducing the number of field employees, cutting back discretionary spending, or maintaining fewer field vehicles.

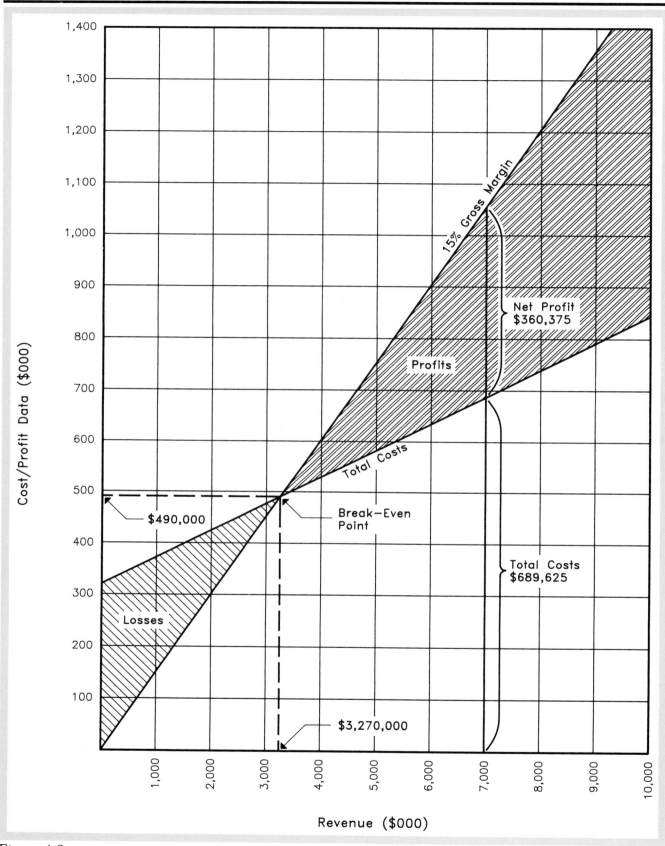

Figure 4.8

5. Change the product "mix" by concentrating on more profitable construction projects. For example, a firm involved in both new construction and renovation may find new construction to be more profitable. Therefore, the greater the portion of volume generated by new commercial work, the greater the profitability and the lower the break-even point.

Contribution Margin

An important concept associated with break-even analysis is that of contribution margin. In manufacturing or retail operations, "contribution margin" refers to the difference between the revenue generated by each unit of product and the variable costs incurred in producing each unit. This difference is the sum available to offset, or contribute to, fixed costs and, ultimately, a profit. However, since contractors do not produce standardized product units, their contribution margin is best thought of as the difference between gross profit dollars and the variable operating costs associated with each such dollar.

The contribution margin (CM) for contractors can be determined using the following formula.

$$\text{Contribution Margin (CM)} = \frac{\text{Gross Profit} - (\text{Variable Cost \& Var. Portion Semi-Variable Cost})}{\text{Gross Profit}}$$

Using the data from our previous examples at the $7 million sales level, we have:

$$CM = \frac{\$1,050,000 - (\$161,875 + \$209,250)}{\$1,050,000} = 0.6465$$

This means that for every dollar of gross profit earned, approximately 65% is left over after covering all variable costs and the variable portion of semi-variable costs. This sum is available to be applied, or contributed, to the payment of fixed costs and profit.

The concept of contribution margin can be used to mathematically compute the break-even volume, as follows:

$$BEV = \frac{FC + FPSVC}{CM} \div GPM$$

where:

BEV = Break-even Volume
GPM = Gross Profit Margin

In our example, break-even volume is calculated as follows:

$$BEV = \frac{(\$106,500 + \$212,000)}{0.65} \div 0.15 = \$3,266,666$$

This figure for break-even volume closely matches that which is derived graphically in Figure 4.8. Eastway must therefore complete approximately $3,270,000 in sales volume, earning a gross margin of 15%, in order to cover the operating costs incurred by the firm at that level of sales. If the firm determines that it will earn a 13% gross margin due to competitive pressures, its new break-even volume must increase if total costs are still to be covered. This new volume is computed as follows:

$$BEV = \frac{(\$106,500 + \$212,000)}{0.65} - 0.13 = \$3,769,230$$

The additional sales—approximately $500,000—is the amount required to offset a drop in gross margin from 15% to 13%.

The break-even relationship can be further expanded to determine the volume required to generate a specified level of net profit. We simply add the desired profit to fixed costs and proceed with the computation as follows:

$$\text{Required volume} = \frac{FC + FPSVC + NP}{CM} \div GPM$$

where NP = Net Profit

If our goal is a net profit of $360,375, as earned in our example, the computation would be as follows:

$$\text{Required volume} =$$
$$\frac{(\$106,500 + \$212,000 + 360,375)}{0.65} \div 0.15 = \$6,962,820$$

Again, allowing for a small rounding error, this figure closely matches the $7 million dollar volume reflected in Figure 4.8, yielding a profit of $360,375.

Sample Operating Budget

Many of the cost and financial forecasting concepts covered thus far in this chapter are illustrated and expanded in the following discussion of Eastway Contractors' 1988 Operating Budget. While a single, consolidated budget will serve as our example, larger firms may, in fact, develop into multi-division organizations with several profit or *responsibility centers*. These centers may be geographical or functional in nature, with separate operating budgets for each. Having an accounting of profit or responsibility centers allows management to track and evaluate the performance of each center. For the sake of simplicity, we will view Eastway as a single profit center. The firm's job cost system should, however, be utilized as a form of responsibility accounting for each building project. In a sense, each project supervisor is then held accountable for the controllable costs on his respective project(s), and is thereby considered a responsibility center.

The budget shown in Figure 4.9 reflects the projected results of many assumptions and data (see Figure 4.2) input into an electronic spreadsheet template. The template is designed to process such data and produce multiple flexible budgets based on various "what if" assumptions pertaining to different levels of sales, gross margins, and costs. The figures shown for *Indirect* costs and *General and Administrative* costs are supported by the schedules shown in Figures 4.10 and 4.11. These schedules are designed to automatically recalculate all individual costs at various levels of sales and then to feed the summarized results into the Budget Summary shown in Figure 4.9.

Based on the projections outlined in the marketing plan, the president of Eastway believes the firm will complete $8 million in sales in 1988. It is expected the firm will realize a 14% gross margin at this volume, which is slightly lower than the industry median (due to local competitive pressure). The president recognizes that when the firm exceeds $7 million in sales, a number of fixed costs will increase in order to accommodate the new level of sales. Fortunately, the new level of fixed costs so incurred will remain constant from $7 million to $10 million in sales—the new relevant range for the firm's budgeting purposes.

EASTWAY CONTRACTORS, INC.
1988 BUDGET SUMMARY
RELEVANT RANGE $7-$10 MILLION

DIRECT COST ALLOCATION

	% OF REVENUE	MARKUP	AMOUNT	MATERIAL %	MATERIAL AMOUNT	LABOR %	LABOR AMOUNT	SUBCONTRACT %	SUBCONTRACT AMOUNT	EQUIPMENT %	EQUIPMENT AMOUNT	MISCELLANEOUS %	MISCELLANEOUS AMOUNT
1 PROJECTED REVENUE			$8,000,000										
2 DIRECT COST	86.00%	100.00%	6,880,000	20%	1,376,000	3%	206,400	75%	5,160,000	1%	68,800	1%	68,800
3 GROSS PROFIT	14.00%	16.28%	1,120,000										
4 OPERATING COSTS													

INDIRECT COST ALLOCATION

	% OF REVENUE	MARKUP	AMOUNT	MATERIAL %	MATERIAL AMOUNT	LABOR %	LABOR AMOUNT	SUBCONTRACT %	SUBCONTRACT AMOUNT	EQUIPMENT %	EQUIPMENT AMOUNT	MISCELLANEOUS %	MISCELLANEOUS AMOUNT
5 INDIRECT COSTS													
6 Field Employee Benefits			47,500	20%	9,500	20%	9,500	40%	19,000	10%	4,750	10%	4,750
7 Field Equipment			24,929	20%	4,986	50%	12,464	10%	2,493	10%	2,493	10%	2,493
8 Field Insurance			34,000	20%	6,800	3%	1,020	75%	25,500	1%	340	1%	340
9 Field Vehicles			29,000	20%	5,800	30%	8,700	40%	11,600	5%	1,450	5%	1,450
10 Shop Maintenance			6,000	30%	1,800	40%	2,400	0%	0	20%	1,200	10%	600
11 Warranty			24,000	20%	4,800	3%	720	75%	18,000	1%	240	1%	240
12 Misc. Indirect Expenses			16,000	20%	3,200	3%	480	75%	12,000	1%	160	1%	160
13 TOTAL INDIRECT COSTS	2.27%	2.64%	181,429	2.68%	36,886	17.10%	35,284	1.72%	88,593	15.45%	10,633	14.58%	10,033

	% OF REVENUE	MARKUP	AMOUNT
14 GENERAL & ADMINISTRATIVE COSTS			
15 Business Insurance			10,000
16 Occupancy			56,857
17 Office Employees			200,357
18 Office Equipment			11,700
19 Office Supplies			15,000
20 Office Vehicles			19,757
21 Officer Compensation			189,500
22 Professional Fees			15,714
23 Sales & Marketing			47,000
24 Travel & Entertainment			14,286
25 Misc. Gen. & Admin. Expenses			45,429
26 TOTAL GEN. & ADMIN. COSTS	7.82%	9.09%	625,600
27 TOTAL OPERATING COSTS	10.09%	11.73%	807,029
28 OPERATING PROFIT	3.91%	4.55%	312,971
29 ALL OTHER INCOME & EXPENSE (NET)	0.10%	0.12%	8,000
30 NET PROFIT BEFORE TAX	4.01%	4.67%	$320,971

Figure 4.9

SCHEDULE OF INDIRECT COSTS
1988 BUDGET

		BASE VOLUME EXPENSES	PROJECTED VOLUME EXPENSES	FIXED EXPENSE	VARIABLE EXPENSE	SEMI-VARIABLE EXPENSE FIXED PORTION	VARIABLE PORTION
31	FIELD EMPLOYEE BENEFITS						
32	Holidays	$1,875	$2,000			$1,000	$1,000
33	Insurance - Health	6,500	7,000			3,000	4,000
34	Retirement	6,000	6,500			2,500	4,000
35	Salaries-Bonus	17,500	20,000		20,000		
36	Salaries-Vacation, Sick, Other	9,250	10,000			4,000	6,000
37	Travel Reimbursement	1,750	2,000		2,000		
38	TOTAL FIELD EMPLOYEE EXPENSES	42,875	47,500	0	22,000	10,500	15,000
39	FIELD EQUIPMENT EXPENSES						
40	Depreciation	5,000	5,000	5,000			
41	Insurance	2,500	2,500	2,500			
42	Repairs and Maintenance	5,000	5,429			2,000	3,429
43	Small Tools and Supplies	10,500	12,000		12,000		
44	TOTAL FIELD EQUIPMENT EXPENSES	23,000	24,929	7,500	12,000	2,000	3,429
45	FIELD INSURANCE EXPENSES						
46	Builders' Risk	20,125	23,000		23,000		
47	General Liability	11,000	11,000	11,000			
48	TOTAL FIELD INSURANCE EXPENSES	31,125	34,000	11,000	23,000	0	0
49	FIELD VEHICLE EXPENSES						
50	Depreciation	9,000	9,000	9,000			
51	Fuel & Oil	7,000	8,000		8,000		
52	Insurance	4,000	4,000	4,000			
53	Repairs & Maintenance	7,000	8,000		8,000		
54	TOTAL FIELD VEHICLE EXPENSES	27,000	29,000	13,000	16,000	0	0
55	SHOP MAINTENANCE EXPENSE	5,250	6,000		6,000		
56	WARRANTY EXPENSE	21,000	24,000		24,000		
57	MISCELLANEOUS INDIRECT EXPENSES	14,000	16,000		16,000		
58	TOTAL INDIRECT COSTS	$164,250	$181,429	$31,500	$119,000	$12,500	$18,429

Figure 4.10

112

SCHEDULE OF GENERAL & ADMINISTRATIVE EXPENSES
1988 BUDGET

	BASE VOLUME EXPENSES	PROJECTED VOLUME EXPENSES	FIXED EXPENSE	VARIABLE EXPENSE	SEMI-VARIABLE EXPENSE	
					FIXED PORTION	VARIABLE PORTION
59 BUSINESS INSURANCE						
60 Building	$2,000	$2,000	$2,000			
61 General Liability & Office Equipment	8,000	8,000	8,000			
62 TOTAL BUSINESS INSURANCE	10,000	10,000	10,000	0	0	0
63 OCCUPANCY EXPENSES						
64 Rent	36,000	36,000	36,000			
65 Repairs & Maintenance	6,000	6,000	6,000			
66 Taxes	3,000	3,000	3,000			
67 Telephone	7,500	7,857			5,000	2,857
68 Utilities	4,000	4,000	4,000			
69 TOTAL OCCUPANCY EXPENSES	56,500	56,857	49,000	0	5,000	2,857
70 OFFICE EMPLOYEE EXPENSES						
71 Continuing Education	4,625	5,000			2,000	3,000
72 Insurance - Health	6,125	6,500			3,500	3,000
73 Insurance - Workers Comp.	1,875	2,000			1,000	1,000
74 Payroll Taxes	16,750	18,000			8,000	10,000
75 Retirement	5,625	6,000			3,000	3,000
76 Salaries-Bonus	14,000	16,000		16,000		
77 Salaries-Staff	135,000	142,857			80,000	62,857
78 Social Activities	3,750	4,000			2,000	2,000
79 TOTAL OFFICE EMPLOYEE EXPENSES	187,750	200,357	0	16,000	99,500	84,857
80 OFFICE EQUIPMENT EXPENSES						
81 Depreciation	7,500	7,500	7,500			
82 Rental	1,750	2,000		2,000		
83 Repairs	2,025	2,200			800	1,400
84 TOTAL OFFICE EQUIPMENT EXPENSES	11,275	11,700	7,500	2,000	800	1,400
85 OFFICE SUPPLIES	14,000	15,000			7,000	8,000
86 OFFICE VEHICLE EXPENSES						
87 Depreciation	12,000	12,000	12,000			
88 Fuel & Oil	3,000	3,257			1,200	2,057
89 Insurance	2,500	2,500	2,500			
90 Repairs & Maintenance	2,000	2,000	2,000			
91 TOTAL OFFICE VEHICLE EXPENSES	19,500	19,757	16,500	0	1,200	2,057
92 OFFICERS COMPENSATION						
93 Insurance-Disability	2,500	2,500	2,500			
94 Insurance-Life	3,000	3,000	3,000			
95 Salaries-Officers	167,500	180,000			80,000	100,000
96 Retirement	4,000	4,000	4,000			
97 TOTAL OFFICERS COMPENSATION	177,000	189,500	9,500	0	80,000	100,000

Figure 4.11

98	PROFESSIONAL FEES					
99	Accounting	7,000	7,286			5,000	2,286
100	Legal	5,000	5,286			3,000	2,286
101	Other	3,000	3,143			2,000	1,143
102	TOTAL PROFESSIONAL FEES	15,000	15,714	0	0	10,000	5,714
103	SALES & MARKETING EXPENSES					
104	Advertising	7,500	8,000			4,000	4,000
105	Commissions	26,250	30,000		30,000		
106	Literature	2,000	2,000	2,000			
107	Promotions	6,500	7,000			3,000	4,000
108	TOTAL SALES & MARKETING EXPENSES	42,250	47,000	2,000	30,000	7,000	8,000
109	TRAVEL & ENTERTAINMENT EXPENSES					
110	Meals	3,500	3,714			2,000	1,714
111	Travel	10,000	10,571			6,000	4,571
112	TOTAL TRAVEL & ENTERTAINMENT EXPENSES	13,500	14,286	0	0	8,000	6,286
113	OTHER GEN. & ADMIN. EXPENSES					
114	Bad Debt	10,500	12,000		12,000		
115	Classified Advertising	5,500	6,000			2,000	4,000
116	Contributions	9,000	9,429			6,000	3,429
117	Dues & Subscriptions	2,500	2,500	2,500			
118	Fines & Penalties	1,750	2,000		2,000		
119	Licensing	1,500	1,500	1,500			
120	Temporary Secretarial	3,500	4,000		4,000		
121	Miscellaneous	7,000	8,000		8,000		
122	TOTAL OTHER GEN. & ADMIN. EXPENSES	41,250	45,429	4,000	26,000	8,000	7,429
123	TOTAL GEN. & ADMIN. EXPENSES	$588,025	$625,600	$98,500	$74,000	$226,500	$226,600
124	OTHER INCOME & EXPENSE					
125	OTHER INCOME					
126	Interest Income	($11,000)	($12,000)			($4,000)	(8,000)
127	Miscellaneous Income	(7,000)	(8,000)		(8,000)		
128	OTHER EXPENSE					
129	Interest Expense	11,500	12,000			8,000	4,000
130	TOTAL OTHER INCOME & EXPENSE (NET)	($6,500)	($8,000)	$0	($8,000)	$4,000	($4,000)
131	TOTAL ALL EXPENSES	$745,775	$799,029	$130,000	$185,000	243,000	241,029
						TOTAL SEMI-VARIABLE	$484,029

Figure 4.11 (*continued*)

114

The controller uses actual cost data from 1987 (when the firm had $6 million in sales) to project increases in fixed costs and the fixed portion of semi-variable costs that will be incurred when the firm moves up to, but not beyond, $7 million in sales. The results of this projection are shown in Figure 4.3, which reflects the firm's operating results given the current relevant range from $4 million to $7 million.

Using the complete budget at the upper end of the previous relevant range, $7 million, the controller then makes a number of adjustments to fixed costs and the fixed portion of semi-variable costs. These adjustments are based on a multitude of assumptions and hard data, as shown in Figure 4.2. All such adjustments are intended to establish the expected results of operations within the new relevant range, in this case from $7 million to $10 million in annual sales.

The following sections explain the format of the budget template shown in Figures 4.9–4.11. Also explained are the various cost adjustments made by Eastway's controller. This is all part of the process of establishing the new level of fixed costs and the fixed portions of semi-variable costs—within the relevant range from $7 million to $10 million in sales.

Budget Summary

Line 1 of Figure 4.9 indicates that revenue for 1988 is projected at $8,000,000. The remainder of the budget summary reflects the total operating costs at this level of sales, with net profit dependent on the gross margin that management believes the firm can earn.

Direct Cost: Line 2 (Figure 4.9) indicates that the firm expects to incur direct costs (labor, materials, subcontractors, equipment, and miscellaneous) of $6,880,000 in producing sales of $8,000,000. This sum is based on the actual project estimates for both projects under contract and those under negotiation that are expected to be started within the year. Under the *Percent of Revenue* column, $6,880,000 represents 86% of the projected revenue of $8,000,000.

The figures listed under the *Mark-up* column represent the percentage of direct cost committed to specific cost or profit elements. In all cases, the Percent of Revenue will be different from the mark-up. For example, total operating costs of $807,029 represent 10.09% of the $8 million projected revenue, but 11.73% of the $6,880,000 direct cost. If a manager expected revenue to equal $8 million, and operating costs to equal 10.09% of revenue, he could easily determine the total dollars committed to operating costs by multiplying as follows:

$8,000,000 × 0.1009 = $807,200

This $807,200 closely matches (allowing for a small rounding error) the $807,029 reflected in the budget.

On the other hand, the manager may anticipate direct costs in 1988 of $6.88 million, and operating costs equal to 11.73% of direct costs. Using these figures, he could determine the total mark-up on direct costs required to cover the firm's operating costs. This figure can be calculated by multiplying as follows:

$6,880,000 × 0.1173 = $807,024

Again, the result is slightly different from the budget figure of $807,029 due to a minor rounding error.

The direct cost of $6.88 million anticipated by Eastway in 1988 is further allocated in line 2 to material, labor, subcontract, equipment, and miscellaneous. This distribution of direct costs is determined by reviewing the cost summaries for the individual projects. Because Eastway has intentionally favored the use of subcontractors (rather than internal work forces and direct purchase of materials), 75% of total direct costs, or $5.16 million, is allocated to subcontractors. Material purchases made directly by the firm account for 20%; labor provided directly by the firm accounts for 3%; and only 1% of direct costs are attributed to both equipment and miscellaneous.

Gross Profit: Line 3 indicates that the firm's gross profit (revenue less direct costs) is projected at $1.12 million. This figure represents 14% of total revenue, the sum available to defray all operating costs and yield a profit.

Operating Costs: Lines 4 through 27 include all data pertaining to both the *Indirect* costs and *General and Administrative* costs that management expects the firm to incur at a volume of $8 million. The individual line items are all summarized figures from the supporting schedules shown in Figures 4.10, 4.11, and 4.12. As shown in line 13, *Total Indirect* costs of $181,429 represent 2.27% of revenue, while line 26 indicates that *Total General and Administrative* costs of $625,600 represent 7.82% of revenue. The *Total Operating* costs of $807,029 shown in line 27, equal 10.09% (2.27% + 7.82%).

The *Indirect* costs shown in lines 6–12 are also allocated amongst the direct cost components. For example, line 7 projects a total cost of $24,929 for *Field Equipment* expenses. These are expenses incurred by the firm for its own field equipment, as detailed in lines 40 through 43 in Figure 4.10. These costs result from equipment usage on a number of projects, and cannot be accurately or conveniently job-costed to individual jobs. As shown, the managers of the firm have determined that 20% of these expenses will be incurred as the result of material handling, 50% to support internal labor, 10% for subcontractors, 10% for equipment use on individual projects, and 10% for miscellaneous costs. Management must analyze each of the indirect costs in terms of the direct cost element that is most responsible.

Another example of an indirect cost is *Shop Maintenance*, which totals $6,000. In Eastway's case, 30% of this expense is for stocking and maintenance of materials in the shop. Another 40% is attributed to labor, for the materials and equipment stored in the shop are primarily intended to support the firm's internal labor force. Zero percent of this item is allocated to subcontractors, for no shop maintenance expenses are incurred in support of or as the result of subcontractors. Equipment is responsible for 20% of this cost, for the storage, cleaning, and maintenance of equipment represents $1200 of the total cost. The final 10% has been allocated to *Miscellaneous*, for a total of 100%.

In some cases, this allocation process requires a high degree of "imagineering", for it is not always a clear-cut decision. Management must, however, make its best effort to allocate each indirect cost, such that the totals will be as accurate as possible. The key is having the totals in line 13 yield mark-up percentages that, when multiplied by the direct cost elements for any given job, will cover the indirect costs associated with the individual direct cost elements for that specific job. In the case of Eastway, for example, the total indirect costs for subcontractors is $88,593, or 1.72% of the total direct costs of $5,160,000 expended for subcontractors. Total indirect costs for material is $36,886, which represents 2.68% of the $1,376,000 expended for materials.

Suppose, for example, that Eastway were to bid $1,162,800 for the construction of Project X, with total direct costs of $1 million (86% of revenue). Assuming the direct cost element distribution as shown below, the indirect costs allocated to the project would be determined as follows:

Cost Element	Direct Cost	Indirect Mark-up	Allocated Cost
Materials	200,000 ×	2.68% =	$5,360
Labor	30,000 ×	17.10% =	5,130
Subcontractors	750,000 ×	1.72% =	12,900
Equipment	10,000 ×	15.45% =	1,545
Miscellaneous	10,000 ×	14.58% =	1,458
Total Direct Costs	**$1,000,000**		**$26,393**

A total of $26,393 must be included in the total bid to cover the indirect costs resulting from Project X. It is interesting to note that this sum does, in fact, closely equal the mark-up of 2.64% suggested in line 13 under the Mark-up column.

Lines 14 through 26 relate to the general and administrative (G & A) costs that management expects to incur when operating at a sales level of $8 million. In this example, total G & A costs are $625,600, or 7.82% of revenue. As a mark-up on direct cost, this sum represents 9.09% of $6,880,000. Using the previous example, G & A allocated to Project X can be computed as follows:

Direct Cost $1,000,000 × 9.09% = $90,900

Total direct costs and operating expenses (excluding profit) incurred by Eastway in the execution of Project X would be computed as follows:

Direct Costs	$1,000,000
Indirect Costs	26,393
General and Administrative Costs	90,900
Total Direct Costs & Operating Expenses	**$1,117,293**

Total Operating Costs: Line 27 reflects the sum of indirect and G & A costs, in this case $807,029. This represents 10.09% of revenue (2.27% for indirect costs plus 7.82% for G & A costs), or 11.73% of direct costs. Therefore, management knows it must mark up its direct costs on its projects in 1988 by an *average* of 11.73% in order to cover the firm's operating costs.

Operating Profit: Given a sales volume of $8 million, a 14% gross profit, and operating costs of $807,029 as projected in this example, Eastway will earn an operating profit of $312,971. This sum is referred to as *operating profit*, as it includes only the income and expenses that are related to the ongoing operations of the firm (excluding *All Other Income & Expense*).

All Other Income and Expense (NET): Line 29 is a net figure, in this case $8,000, that includes miscellaneous other income and expenses that are not generated through the ongoing operations of the firm. Examples include interest expense, interest income, dividends received, and gains or losses on the sale of assets.

Net Profit Before Tax: Given an operating profit of $312,971 and $8,000 in miscellaneous income, the firm will earn a total of $320,971 in 1988. This number will be further adjusted based on the firm's tax liability, which may vary significantly due to a number of factors that will be covered in Chapter 7, "Tax Management".

Schedule of Indirect Costs

Figure 4.10 is a detailed breakdown of the summarized numbers shown in lines 6–12 of the Budget Summary. For instance, Line 6 in the Budget Summary indicates an expenditure of $47,500 for *Field Employee Benefits*. This sum is supported by lines 32–37 in the Schedule of Indirect Costs, which total $47,500 under the column entitled *Projected Volume Expenses*. All indirect costs and G & A costs summarized in the Budget Summary (Figure 4.9) are supported by the detail shown in Figures 4.10, 4.11 and 4.12. Formulas are built into the spreadsheet template to automatically compute these detailed costs at various levels of volume and to feed the results into the Budget Summary. The schedules for indirect costs and G & A costs have six columns of figures, each explained in the following sections.

Base Volume Costs

The figures in this column represent the expenses that management expects the firm to incur at the low end of the relevant range, in this case $7 million.

Projected Volume Costs

This column lists the projected expenses to be incurred at a specified volume above the Base Volume, in this case $8 million. These figures are all appropriately adjusted for changes in variable and semi-variable costs due to the increase in volume.

Fixed Costs

Individual costs that are fixed within the given relevant range are listed under this column.

Variable Costs

Costs that vary in the same direction as sales, and in equal proportion, are listed in this column.

Semi-Variable Costs

The *Semi-variable Costs* column is further divided into two columns entitled *Fixed Portion* and *Variable Portion*. Some costs have both a fixed component and a variable component. The column entitled *Fixed Portion* lists the portion of semi-variable costs that are fixed within the relevant range. The figures represent the minimum expenditure that the firm must make in any given cost category while still maintaining the firm in an operational condition. The *Variable Portion* is that part of any semi-variable cost that *exceeds* the fixed portion.

Lines 39–44 provide detailed information pertaining to *Field Equipment Expenses*. Depreciation of $5,000 is listed as a fixed cost, for it will remain the same whether the firm completes $7 million or $10 million in sales. In other words, no equipment must be purchased or sold within this range, so the depreciation associated with this equipment is constant. For the same reasons, the insurance costs of $2,500 related to field equipment are fixed.

Repairs and Maintenance costs at a volume of $8 million are $5,429, and are listed as semi-variable. At a base volume of $7 million, management believes that the firm will incur $5,000 for repairs and maintenance of field equipment. $2,000 of this figure is listed as *fixed*, for management feels this is the minimum sum that the firm must expend for repairs and maintenance (regardless of volume) if it is to be fully operational within the relevant range. In other words, even in the most extreme condition of *no sales*, the firm still must spend $2,000 on repairs and maintenance in order to be ready to produce construction valued within the relevant range.

The remainder of this $5,000 cost, $3,000, is considered the variable portion of this semi-variable cost. As sales increase, this $3,000 cost will rise proportionately. In this case, the base volume is $7 million, while the projected volume is $8 million, a 14.28% increase. Likewise, the $3,000 base volume variable cost for Repairs and Maintenance is shown as $3,429, also a 14.28% increase.

When Eastway increases its relevant range from $4–$7 million up to $7–$10 million, virtually all fixed indirect costs also increase. This is because the quantity of field equipment and vehicles must make a "lumpy", stepped jump when sales are increased beyond the $7 million level. The fixed depreciation and insurance costs associated with these fixed assets must, therefore, also increase. On the other hand, the fixed portion of semi-variable costs for field employee benefits remains the same, while the variable portion at the base volume increases with the additional work force required.

Line 58 contains the totals for all six columns. It is found that indirect costs of $164,250 at the $7 million base volume are increased to $181,429 at the $8 million level due to the upward adjustment in variable costs and the variable portion of semi-variable costs. It should be noted that this $181,429 is the same as that shown in line 13 of the Budget Summary.

Schedule of General and Administrative Costs

Figures 4.11 and 4.12 portray the G & A costs in the same format as the indirect costs that were just explained. Again, many of the base volume fixed costs for G & A within the new relevant range are higher due to the higher volume. For instance, fixed occupancy expenses all increase because of the need for expanded office space. Office vehicle expense is adjusted upward due to the purchase of one additional vehicle. Likewise, many of the fixed portions of semi-variable costs take a jump at the base volume. Other examples include professional fees, travel and entertainment, and salaries for office staff.

Line 123 (Figure 4.12) shows the totals for all G & A costs by cost element. The total of $625,600 at the $8 million sales level matches that shown on Line 26 in Figure 4.9.

Lines 124–130 (Figure 4.12) relate to *Other Income and Expense (NET)*. Line 126 shows interest income of ($12,000) under the *Projected Volume Cost* column. Because this sum is income, rather than an expense, it is shown in parentheses as a negative number. The interest expense of $12,000 shown in line 129 is, on the other hand, a positive figure because it is an expense, not income. The total ($8,000) in line 130 is added to line 28, *Operating Profit*.

All costs incurred by the firm at a volume of $8 million, broken down by cost elements, is shown in line 131. The $745,775 in costs incurred at the $7 million level is increased to $799,029 at the $8 million level, due to the increase in variable costs and the variable portion of semi-variable costs.

Net Income Forecast

When the spreadsheet template for budgeting has been created, with all required formulas and base volume cost data in place, the controller has only to adjust the projected volume and gross profit margin to determine the firm's performance under any number of scenarios. Figure 4.12 is a summarized Net Income Forecast that projects costs and profit in the relevant range from $7–$10 million, and with gross margins of 13%–16%. This forecast is invaluable in guiding management in the formulation of mark-ups on any given project during the course of the year.

One final step in the forecasting process is graphing the results. Figure 4.13 depicts the firm's performance over the continuum of sales levels from $0 to $10 million. The totals in line 131 of Figure 4.11 are shown in this graph at the $8 million projected level. It is interesting to note that the break-even point has shifted to the right of the previous break-even point, due to the increase in fixed costs incurred within the new relevant range.

```
                               EASTWAY CONTRACTORS, INC.
                                 NET INCOME FORECAST
                               RELEVANT RANGE $7M - $10M

--------------------------------------------------------------------------------------------------
     FINANCIAL     - - - - - - - - - - - - - GROSS PROFIT % - - - - - - - - - - - - - - - - - -
       DATA          13%                 14%                 15%                 16%
--------------------------------------------------------------------------------------------------
          REVENUE  $7,000,000  100.00%  $7,000,000  100.00%  $7,000,000  100.00%  $7,000,000  100.00%
      DIRECT COST  (6,090,000)  -87.00%  (6,020,000)  -86.00%  (5,950,000)  -85.00%  (5,880,000)  -84.00%
                   ----------  -------  ----------  -------  ----------  -------  ----------  --------
     GROSS PROFIT     910,000   13.00%     980,000   14.00%   1,050,000   15.00%   1,120,000   16.00%
                   ----------  -------  ----------  -------  ----------  -------  ----------  --------
      FIXED COSTS    (130,000)   -1.86%    (130,000)   -1.86%    (130,000)   -1.86%    (130,000)   -1.86%
 FIXED PORT. S. V. COSTS (243,000) -3.47% (243,000)   -3.47%    (243,000)   -3.47%    (243,000)   -3.47%
  VAR. PORT. S. V. COSTS (210,900) -3.01% (210,900)   -3.01%    (210,900)   -3.01%    (210,900)   -3.01%
   VARIABLE COSTS    (161,875)   -2.31%    (161,875)   -2.31%    (161,875)   -2.31%    (161,875)   -2.31%
                   ----------  -------  ----------  -------  ----------  -------  ----------  --------
NET PROFIT BEFORE TAX $164,225   2.35%    $234,225    3.35%    $304,225    4.35%    $374,225    5.35%
                   ==========  =======  ==========  =======  ==========  =======  ==========  ========

          REVENUE  $8,000,000  100.00%  $8,000,000  100.00%  $8,000,000  100.00%  $8,000,000  100.00%
      DIRECT COST  (6,960,000)  -87.00%  (6,880,000)  -86.00%  (6,800,000)  -85.00%  (6,720,000)  -84.00%
                   ----------  -------  ----------  -------  ----------  -------  ----------  --------
     GROSS PROFIT   1,040,000   13.00%   1,120,000   14.00%   1,200,000   15.00%   1,280,000   16.00%
                   ----------  -------  ----------  -------  ----------  -------  ----------  --------
      FIXED COSTS    (130,000)   -1.63%    (130,000)   -1.63%    (130,000)   -1.63%    (130,000)   -1.63%
 FIXED PORT. S. V. COSTS (243,000) -3.04% (243,000)   -3.04%    (243,000)   -3.04%    (243,000)   -3.04%
  VAR. PORT. S. V. COSTS (241,029) -3.01% (241,029)   -3.01%    (241,029)   -3.01%    (241,029)   -3.01%
   VARIABLE COSTS    (185,000)   -2.31%    (185,000)   -2.31%    (185,000)   -2.31%    (185,000)   -2.31%
                   ----------  -------  ----------  -------  ----------  -------  ----------  --------
NET PROFIT BEFORE TAX $240,971   3.01%    $320,971    4.01%    $400,971    5.01%    $480,971    6.01%
                   ==========  =======  ==========  =======  ==========  =======  ==========  ========

          REVENUE  $9,000,000  100.00%  $9,000,000  100.00%  $9,000,000  100.00%  $9,000,000  100.00%
      DIRECT COST  (7,830,000)  -87.00%  (7,740,000)  -86.00%  (7,650,000)  -85.00%  (7,560,000)  -84.00%
                   ----------  -------  ----------  -------  ----------  -------  ----------  --------
     GROSS PROFIT   1,170,000   13.00%   1,260,000   14.00%   1,350,000   15.00%   1,440,000   16.00%
                   ----------  -------  ----------  -------  ----------  -------  ----------  --------
      FIXED COSTS    (130,000)   -1.44%    (130,000)   -1.44%    (130,000)   -1.44%    (130,000)   -1.44%
 FIXED PORT. S. V. COSTS (243,000) -2.70% (243,000)   -2.70%    (243,000)   -2.70%    (243,000)   -2.70%
  VAR. PORT. S. V. COSTS (271,157) -3.01% (271,157)   -3.01%    (271,157)   -3.01%    (271,157)   -3.01%
   VARIABLE COSTS    (208,125)   -2.31%    (208,125)   -2.31%    (208,125)   -2.31%    (208,125)   -2.31%
                   ----------  -------  ----------  -------  ----------  -------  ----------  --------
NET PROFIT BEFORE TAX $317,718   3.53%    $407,718    4.53%    $497,718    5.53%    $587,718    6.53%
                   ==========  =======  ==========  =======  ==========  =======  ==========  ========

          REVENUE  $10,000,000 100.00%  $10,000,000 100.00%  $10,000,000 100.00%  $10,000,000 100.00%
      DIRECT COST  (8,700,000)  -87.00%  (8,600,000)  -86.00%  (8,500,000)  -85.00%  (8,400,000)  -84.00%
                   ----------  -------  ----------  -------  ----------  -------  ----------  --------
     GROSS PROFIT   1,300,000   13.00%   1,400,000   14.00%   1,500,000   15.00%   1,600,000   16.00%
                   ----------  -------  ----------  -------  ----------  -------  ----------  --------
      FIXED COSTS    (130,000)   -1.30%    (130,000)   -1.30%    (130,000)   -1.30%    (130,000)   -1.30%
 FIXED PORT. S. V. COSTS (243,000) -2.43% (243,000)   -2.43%    (243,000)   -2.43%    (243,000)   -2.43%
  VAR. PORT. S. V. COSTS (301,286) -3.01% (301,286)   -3.01%    (301,286)   -3.01%    (301,286)   -3.01%
   VARIABLE COSTS    (231,250)   -2.31%    (231,250)   -2.31%    (231,250)   -2.31%    (231,250)   -2.31%
                   ----------  -------  ----------  -------  ----------  -------  ----------  --------
NET PROFIT BEFORE TAX $394,464   3.94%    $494,464    4.94%    $594,464    5.94%    $694,464    6.94%
                   ==========  =======  ==========  =======  ==========  =======  ==========  ========
```

Figure 4.12

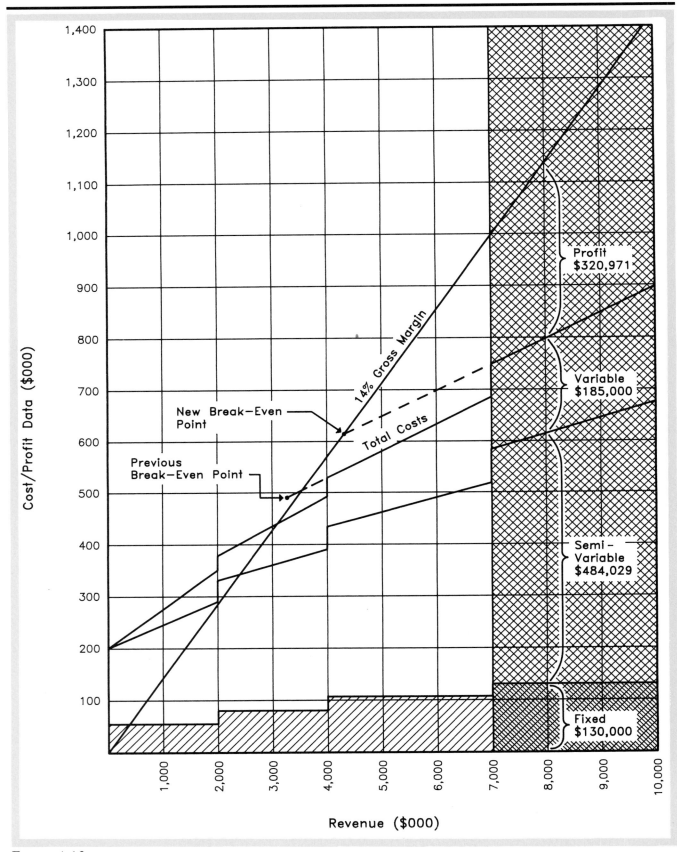

Figure 4.13

Chapter Five

CASH
MANAGEMENT

Chapter Five

CASH MANAGEMENT

Growing construction firms are highly dependent on a level of cash that is adequate to maintain operations. It is the responsibility of top management to ensure this level is maintained. Cash management is, therefore, one of the most important aspects of effective financial control for construction contractors. It is also one of the most neglected aspects.

Unfortunately, it is not unusual for managers to leave their firms' cash position to chance, with little effort expended on planning or managing cash. This shortfall has been one of the primary causes of failures in the industry. Many contractors have also made the mistake of assuming that working capital is a direct indicator of the company's capacity to furnish cash when needed. While this may generally be true over time, working capital is not always a reliable indicator of the ability to meet cash needs at a specific point in time.

Effective cash management maximizes the level of cash within the firm and the return earned from the investment of cash. One of the major goals of cash management is to minimize the time during which cash is not producing income. The five basic cash management principles listed below apply equally to small and large firms.

1. Cash budgeting, which relates to planning the firm's cash needs.
2. Acceleration of cash receipts from customers.
3. Deceleration of cash disbursements to creditors.
4. Raising cash when needed.
5. Short term investment to preserve the value of working capital and maximize earnings.

Successful cash management for the contractor requires that cash as an asset be viewed just like any other income-producing asset. It would not make sense for the contractor to invest in a piece of equipment and then have the equipment sit idle on a job site. Nor does it make sense to acquire equipment without first examining the best possible financing available to pay for the equipment. The same is true for cash, which should not sit idle, nor should it be procured at less than the best possible terms.

The proper level of cash, depending on the size and volume of the firm, must be maintained for four primary reasons:

1. **Transactions.** The contractor *must* maintain a cash balance sufficient to cover the cash needs of the firm. Employees are not interested in receiving IOU's in their pay envelope at the end of the week. Subcontractors and suppliers also expect to be compensated in accordance with their contracts and credit terms.

2. **Compensating balance requirements.** The contractor's banking institution may require that a minimum cash balance be maintained as part of a loan agreement or in order to receive preferential account privileges.

3. **Precautionary reasons.** Since cash flow management is not an exact science, the contractor will want to maintain a cash balance "cushion" to soften the effect of unexpected cash demands.

4. **Investment/Speculation.** Cash should be an income-producing asset. The return that can be generated on a cash investment is highly desirable, such that a portion of this asset should be committed to investment. Accumulated cash can also be used to take advantage of bargain purchase opportunities.

The contractor's cash management system should be geared to minimize the amount of cash required to satisfy the first three of the above listed requirements, all of which are non-income producing. In this way, the amount of cash available for investment is maximized, producing the highest return.

To achieve the goal of minimizing the cash holding requirements stated above, the contractor must enhance his ability to forecast the cash needs of the firm. To use an extreme example, suppose the controller of Eastway Contractors, in the process of writing paychecks one week, discovers that there is only enough cash to pay one half of the employees of the firm. Knowing that this situation may bring about severe consequences, the controller telephones the firm's bank and requests an immediate loan to cover the short term cash shortage. Chances are, the banker will not have a favorable impression of Eastway Contractor's cash management ability. There may be legitimate reasons for this problem, but the fact remains that the controller would be at the mercy of the lending institution and the loan terms it demanded. With better planning, this crisis situation might have been avoided.

The Cash Flow Cycle

In every contracting operation, there is an ongoing cycle of financial transactions resulting from the production and sale of the firm's services. This cycle continuously consumes and replenishes the stock of funds, referred to as *working capital*, available to the firm to finance current operations.

The inflow and outflow of funds rarely balances out to a steady level of available cash. Even if the firm is earning steady profits, there is no guarantee that the level of cash will grow. For example, a series of unusually large subcontractor billings may become due in one month, while the draws against the contract may not become available to the firm until the following month. Any number of special circumstances may cause wild fluctuations in the supply and demand for the firm's cash.

Figure 5.1 illustrates the normal cash flow cycle for the contractor. The arrows directed *into* "cash" indicate *sources* of cash while arrows directed *away* from "cash" indicate cash *uses*. A summary of the cash inflows and outflows is as follows:

Cash Inflows
- Accounts Receivable Collection
- Investment by Owners
- Sale of Fixed Assets
- Debt

Cash Outflows
- Accounts Payable Payments
- Payroll
- Indirect Job Costs
- Operating Expenses
- Debt Repayment
- Fixed Asset Acquisition
- Dividends to Owners

The cash flow cycle shown in Figure 5.1 reflects one of the most critical points of cash management for the contractor. The flow of the diagram indicates that the major source of cash for the contractor, accounts receivable collection, is a result of *billings*, based upon work-in-process. Work-in-process is created after costs have been incurred towards project completion. The contractor must be sure that there is sufficient cash available to cover the costs of ongoing projects as well as general operations *until* the draws from the project catch up with the costs incurred. The cash budget discussed in the next section will identify the peaks and valleys in a contractor's cash position.

Cash Budgeting

The cash budgeting process is a significant planning function involving not only the timing of funds flowing in and out of the firm, but also the specific movement and needs for cash within the firm. Without this planning effort, a firm might unexpectedly incur a major cash shortage that could otherwise have been avoided. Such a shortage could, in turn, bring on a number of problems with creditors, or worse, bankruptcy.

Integrating a cash budgeting effort into the overall management of the firm brings about a number of benefits, including:

- Assurance that sufficient funds will always be available for daily, weekly, or monthly transactional needs. The available funds should allow for taking advantage of all possible vendor discounts for early payment.
- Provision of adequate liquidity for corporate growth.
- Provision of more accurate information regarding short term borrowing needs, allowing ample time to make the required borrowing arrangements.
- More effective planning for the investment of excess working capital, which preserves its purchasing power and contributes to profitability.

The cash flow budget is a plan to which management must commit itself. As with other plans, it is the responsibility of management to direct both external and internal events so that they conform to the plan as closely as possible. Actual results must be compared with the plan, significant variances noted, and corrective action taken.

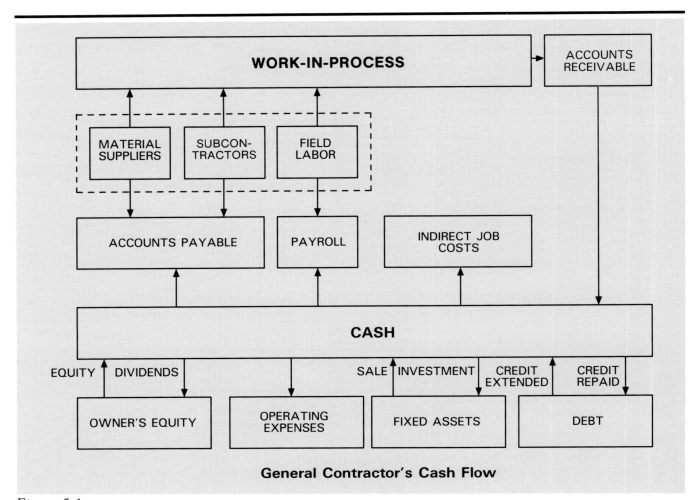

Figure 5.1

Constructing the Cash Budget

There are two major steps involved in the preparation of cash budgets:

1. Development of job cash flow budgets for all work-in-process and other anticipated projects.
2. A consolidated summary cash flow budget for the entire company. The summary budget combines the net cash flows from all individual projects with the operating expenses of the firm.

The Job Cash Flow Budget

The job cash flow budget first requires the contractor to estimate the amounts and timing of cash disbursements and cash receipts related to each project. This information is available from the estimate summary of all direct costs for the project as well as from the project schedule. A study of the schedule, particularly when set up in bar chart form, will reveal the phasing and associated costs of the different trades.

Based on the work expected to be completed during each draw period (typically one month), the manager can refer to the schedule of values against which he is to bill and determine the revenue that should become due each period. Unfortunately, the project schedule and associated draws are directly affected by weather, performance of subcontractors, and a variety of other unknowns. While the contractor may have little control over these factors, the best attempt must nevertheless be made to project conditions as accurately as possible.

A job cash flow budget for a hypothetical job (# 1) is shown in Figure 5.2. Lines 1 to 3 refer to the revenues from progress draws expected each month. The figures should reflect the retainage called for in the contract as well as the time anticipated between the issuance of a draw request and the receipt of funds. As discussed in Chapter 6 on Asset Management, this period of time should be carefully defined in the contract and followed up on diligently.

It should be noted that this particular job cash flow budget is established for a period from January through December. As shown in line 1, the total draws for the entire contract amount to $1,500,000. However, the current year draws are to be $1,000,000, indicating that $500,000 of the total contract was billed in the prior year.

Lines 4 to 9 indicate the direct costs expected to be incurred each month, broken down according to the major cost categories included in the estimate summary. The timing of these disbursements varies with the cost category. For instance, payroll for in-house labor is normally disbursed within one week of incurring the expense; disbursements to suppliers are often required by the end of each month; and disbursements to subcontractors are often within a fixed number of days from the receipt of an invoice. The specific terms arranged with subcontractors and suppliers will certainly affect the timing of disbursements. These variations should all be reflected in the direct cost figures in lines 4 to 9.

With cash receipts and cash disbursements projected over the life of the project, the net cash flow is easily determined, as shown in line 10. The cumulative cash flow found in line 11 is simply the accumulation of cash flows from month to month. A graph depicting the timing of net cash flows for Job # 1 in the current year is shown in Figure 5.3.

EASTWAY CONTRACTORS, INC.
JOB #1 CASH FLOW BUDGET

	ESTIMATE - TOTAL JOB	ESTIMATE - CURRENT YEAR	JANUARY	FEBRUARY	MARCH	APRIL	MAY	JUNE	JULY	AUGUST	SEPTEMBER	OCTOBER	NOVEMBER	DECEMBER
TOTAL JOB REVENUE														
Total Draw	1,500,000	1,000,000	175,000	25,000	100,000	25,000	75,000	150,000	180,000	250,000	20,000			
Retainage		50,000	(17,500)	(2,500)	(10,000)	(2,500)	(7,500)	(15,000)	(18,000)	(25,000)	148,000			
NET RECEIPT	1,500,000	1,050,000	157,500	22,500	90,000	22,500	67,500	135,000	162,000	225,000	168,000	0	0	0
DIRECT JOB COSTS														
Labor	38,000	25,000	2,500	3,500	2,500	3,000	3,500	3,500	3,500	3,500				
Material	271,000	175,000	15,000	15,000	6,000	17,000	23,000	35,000	35,000	30,700				
Subcontractor	950,000	635,000	60,000	72,000	43,000	90,000	40,000	94,000	115,000	118,000				
Equipment	12,000	8,000	0	2,000	2,000	1,000	0	2,000	0	1,000				
Miscellaneous	4,000	2,500	300	300	200	400	400	400	200	500				
TOTAL DIRECT JOB COSTS	1,275,000	845,500	77,800	92,800	53,700	111,400	66,900	134,900	153,700	153,700	0	0	0	0
NET JOB CASH FLOW	225,000	204,500	79,700	(70,300)	36,300	(88,900)	600	100	8,300	71,300	168,000	0	0	0
CUMULATIVE JOB CASH FLOW			100,200	29,900	66,200	(22,700)	(22,100)	(22,000)	(13,700)	57,600	225,600	225,600	225,600	225,600

Figure 5.2

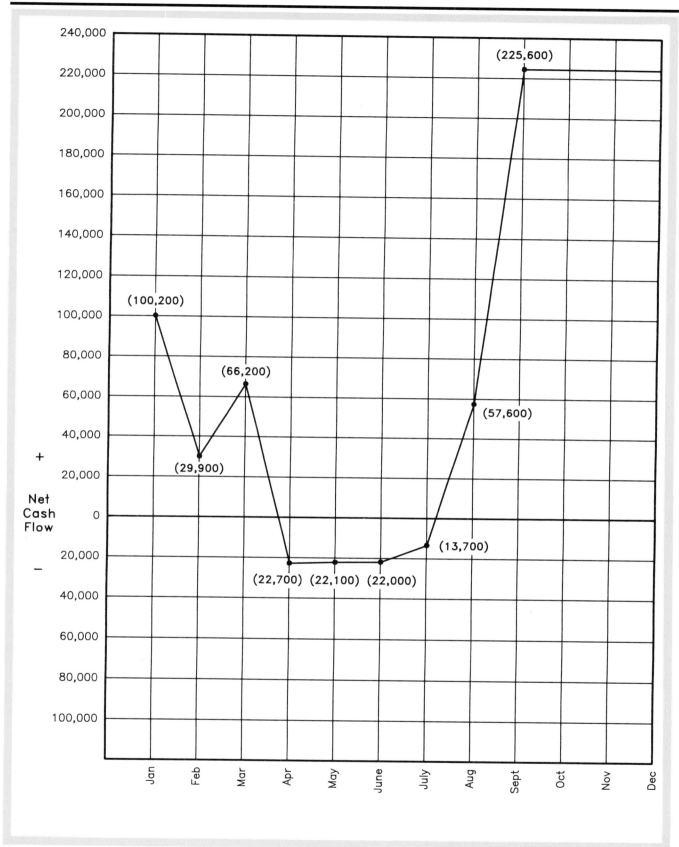

Figure 5.3

131

Figure 5.3 clearly illustrates the kinds of wild fluctuations that can occur in the cash flow of a construction company due to the timing of receipts and disbursements on individual projects. The cash flow on any given project is positive when payments are less than receipts. The cash flow will usually be negative at the start of a project, prior to issuance of any draw requests. It may also be negative when retainage receivable exceeds retainage payable to subcontractors.

It can be seen that by September the project is complete and the cumulative cash flow is $225,600. This means that the firm's gross profit on this $1,500,000 contract is $225,600. This amount is available to contribute to operating expenses incurred by the firm and, hopefully, to a profit.

The Consolidated Company Cash Flow Budget

The company cash flow budget is a compilation of all the individual job cash flow budgets, along with the anticipated operating expenses, capital expenditures, and changes in debt position. This budget is generally prepared for a one-year period and advanced each month with updated information.

The Annual Cash Flow Budget for Eastway Contractors, spanning the period from January 1, 1988 to December 31, 1988 is shown in Figure 5.4. The beginning balance shown in line 1 is simply the cash that the firm has on hand as of January 1. All the information contained in the following lines represents adjustments in any given month that either increase or decrease the beginning balance.

Lines 2 to 10 reflect the net cash flows generated by the various projects (as described in the prior section on job cash flow budgets). A study of Figure 5.2, which is a sample cash flow budget for Job # 1, reveals that the *Net Job Cash Flows* shown in line 10 are, in fact, carried over to the Annual Cash Flow Budget shown in Figure 5.4. This section of the budget should include the net cash flows for both work-in-process and projected new jobs.

The next section of the budget, Figure 5.4, entitled *Expenditures*, covers lines 12 to 18. Lines 12 and 13 represent the monthly anticipated operating expenses, broken down into indirect costs and general and administrative costs. It can be seen that these expenses are spread relatively evenly across the entire year, for operating expenses fluctuate very little for a stable firm with a consistent volume of work. However, special payments for such items as taxes and insurance must be accounted for. Other income and expenses, which are netted against one another each month, are shown in line 14. Line 15 lists any capital expenditures made by the firm. In this case, Eastway plans to purchase a $10,000 computer in the month of September. Line 16 includes the current portion of long term debt, in this case $5,100 for the year, which is distributed fairly evenly across all 12 months. Line 17 is simply a total of all operating expenditures made during the course of the year and for each month.

Line 18 represents the ending cash balance for each month. This is determined by subtracting line 17, *Total Cash Disbursements* for the month, from line 11, *Total Cash Available* each month. This ending cash balance is the sum remaining available to the firm, prior to adjustments for changes in debt.

EASTWAY CONTRACTORS, INC.
ANNUAL CASH FLOW BUDGET
JANUARY 1, 1988 - DECEMBER 31, 1988

	TOTAL	JANUARY	FEBRUARY	MARCH	APRIL	MAY	JUNE	JULY	AUGUST	SEPTEMBER	OCTOBER	NOVEMBER	DECEMBER
BEGINNING BALANCE		85,000	165,900	27,900	29,950	35,600	38,200	31,000	25,800	32,800	77,850	21,800	31,900
NET JOB CASH FLOWS													
In Process:													
Job #1	205,100	79,700	(70,300)	36,300	(88,900)	600	100	8,300	71,300	168,000	0	0	0
Job #2	120,000	120,000	0	0	0	0	0	0	0	0	0	0	0
Job #3	149,500	(48,000)	55,000	48,000	75,000	(22,000)	1,500	40,000	0	0	0	0	0
Job #4	(11,100)	0	(50,000)	25,000	48,000	2,000	500	2,500	(40,000)	(35,000)	20,000	15,000	900
Job #5	(25,300)	(14,000)	(100,000)	15,000	(2,000)	1,500	65,000	(800)	10,000	0	0	0	0
Projected New Jobs:													
Job #6	65,800	0	0	0	0	0	20,000	40,000	(3,000)	(700)	(5,800)	300	15,000
Job #7	82,500	0	(500)	0	55,000	(21,000)	14,000	15,000	400	1,600	10,000	8,000	0
Job #8	30,000	0	0	0	0	0	0	0	0	0	(25,000)	(10,000)	65,000
TOTAL NET JOB CASH FLOWS	616,500	137,700	(165,800)	124,300	87,100	(38,900)	101,100	105,000	38,700	133,900	(800)	13,300	80,900
TOTAL CASH AVAILABLE		222,700	100	152,200	117,050	(3,300)	139,300	136,000	64,500	166,700	77,050	35,100	112,800
EXPENDITURES													
OPERATING EXPENSES													
Indirect Costs	279,000	21,000	23,000	26,000	22,000	23,000	24,000	25,000	24,000	27,000	21,000	20,000	23,000
Gen. & Admin. Costs (less depr.)	414,000	36,000	34,000	36,000	34,000	35,000	34,000	35,000	33,000	35,000	34,000	33,000	35,000
OTHER INCOME & EXPENSES (Net)	(3,100)	(600)	(200)	(150)	50	100	(100)	(200)	(700)	(550)	(250)	(300)	(200)
CAPITAL EXPENDITURES	10,000	0	0	0	0	0	0	0	0	10,000	0	0	0
CURRENT PORTION LONG TERM DEBT	5,100	400	400	400	400	400	400	400	400	400	500	500	500
TOTAL CASH DISBURSEMENTS	705,000	56,800	57,200	62,250	56,450	58,500	58,300	60,200	56,700	71,850	55,250	53,200	58,300
ENDING CASH BALANCE		165,900	(57,100)	89,950	60,600	(61,800)	81,000	75,800	7,800	94,850	21,800	(18,100)	54,500
DEBT ADJUSTMENTS													
LONG TERM DEBT INCREASE (DECREASE)	8,000	0	0	0	0	0	0	0	0	8,000	0	0	0
SHORT TERM DEBT INCREASE (DECREASE)	20,000	0	85,000	(60,000)	(25,000)	100,000	(50,000)	(50,000)	25,000	(25,000)	0	50,000	(30,000)
TOTAL DEBT ADJUSTMENT INC.(DEC.)	28,000	0	85,000	(60,000)	(25,000)	100,000	(50,000)	(50,000)	25,000	(17,000)	0	50,000	(30,000)
ADJUSTED ENDING CASH BALANCE		165,900	27,900	29,950	35,600	38,200	31,000	25,800	32,800	77,850	21,800	31,900	24,500
CUMULATIVE DEBT													
LONG TERM DEBT													
Beginning Balance	40,000	40,000	39,600	39,200	38,800	38,400	38,000	37,600	37,200	36,800	44,400	43,900	43,400
Increase(Decrease)	2,900	(400)	(400)	(400)	(400)	(400)	(400)	(400)	(400)	7,600	(500)	(500)	(500)
ENDING LONG TERM DEBT	42,900	39,600	39,200	38,800	38,400	38,000	37,600	37,200	36,800	44,400	43,900	43,400	42,900
SHORT TERM DEBT													
Beginning Balance	0	0	0	85,000	25,000	0	100,000	50,000	0	25,000	0	0	50,000
Increase(decrease)	20,000	0	85,000	(60,000)	(25,000)	100,000	(50,000)	(50,000)	25,000	(25,000)	0	50,000	(30,000)
ENDING SHORT TERM DEBT	20,000	0	85,000	25,000	0	100,000	50,000	0	25,000	0	0	50,000	20,000

Figure 5.4

133

Lines 19–22 provide information on increases and decreases in long term and short term debt. When the firm borrows funds, debt is increased, and cash available to the firm is also increased. When it repays debt, both debt and available cash are decreased. Line 21 shows the net effect of increases and decreases in short and long term debt. The only change in long term debt was in September, when the firm borrowed $8,000 towards the purchase of a $10,000 computer. Changes in short term debt are quite frequent and primarily due to borrowing and repayments on the firm's line of credit.

The adjusted ending cash balance shown in line 22 is the actual cash available to the firm each month after increases and decreases in debt. The lowest level reached in this balance is $21,800 in October. It can be seen that when the firm does borrow on its line of credit, it normally borrows enough to maintain approximately $25,000–$30,000 cash on hand. Maintaining this amount prevents the need to revisit the bank every week because of unexpected expenditures.

Lines 23 to 28 summarize the long and short term debt for which the firm is responsible on a monthly basis. It should be noted that line 25, *Ending Long Term Debt*, is reduced by $400 each month in response to the monthly payments shown in line 16. This is true until September, when long term debt increases by $7,600 ($8,000 new debt less $400 paid on old debt).

Although the cash budgeting effort can be time consuming, the information provided by an annual cash flow budget (such as that developed for Eastway Contractors) is invaluable. With this information, top management should be able to forecast any short term borrowing needs well in advance. Their primary lender will be quite impressed with the understanding that the firm's managers have of their financial picture, and will be more likely to make the required credit facilities available. With this information, management will be better able to evaluate growth and the associated changes. Furthermore, since they are aware of the future availability of excess working capital, plans can be made as to how this capital might best be invested. These are all important elements in the firm's overall business plan, which points up the critical importance of effective cash budgeting.

Acceleration of Cash Receipts

Ideally, a firm collects cash receipts as soon as possible and delays cash disbursements as long as possible without harming its credit relationships. Accelerating cash receipts should begin with the invoicing process. The earlier the invoices are issued, the earlier receipts can be collected. The contractor should stay up to date on all contracts in order to note the times at which billings are allowable. A schedule of expected billings should be developed and actual billings should be tracked against this schedule. This procedure not only enhances billings, but it helps the contractor monitor jobs that are falling behind. Since each day's delay in billing generally means a day's delay in collection, it is crucial that billings are timely and accurate. Billing terms as specified in the contract should be carefully followed.

The contractor should also establish a strict collection policy and adhere to it. Management tools such as an *aged schedule of accounts receivable* are helpful in maintaining an accounts receivable policy. Chapter 6 covers billing and collection procedures in much more detail.

In addition to timely billings and collections, the contractor can use several other methods to accelerate cash receipts. Among them are:

- Negotiating with the customer for a reduction in the retainage percentage.
- Negotiating with the customer for provisions in the contract which allow the cost of special equipment and/or materials to be billed directly to the customer.
- Front-end loading the schedule of values, such that the project can be over-billed in the early stages of construction.
- Negotiating for early payment on items for which the contractor must make downpayment upon ordering.
- Rapid completion of the job and collection of retainage.

Deposits

Another opportunity for accelerating cash is the prompt deposit of checks. All the efforts exerted in collections are wasted unless the firm makes collected cash available for use as soon as possible. If a check arrives without an invoice or supporting documentation, it should be photocopied and deposited, and researched later. It is good policy to deposit checks as early as possible on the day they are received, since checks deposited in the late afternoon are often not credited to the firm's account until the next day.

Bank Services: When selecting a bank, the contractor should be clear on the bank's policy about making deposited funds available for use. A number of banks impose a waiting period prior to making the deposited funds available. The contractor should avoid banks that impose an unreasonable waiting period. An unreasonable period is one that compares unfavorably with the policies of other banks in the area. It never hurts for the contractor to compare, on occasion, its current bank's policies with those of other banks in the area.

Banks sometimes offer services that help speed up the availability of cash. One of the most common services is the *lock box*. When this service is used, the contractor's customers mail their payments to post office boxes that are emptied several times daily by the firm's bank. The checks are immediately deposited to the contractor's account. The bank then forwards the contractor a list of deposits and any other information required for the contractor's bookkeeping. The major advantage of the lock box is that it cuts down on the time it takes to convert customer checks to collected funds. The primary disadvantages are the cost and the loss of opportunity to review customers' checks prior to deposit. The larger the number of collections and the larger the average receipt, the more likely that a lock box service will be worthwhile.

Float

Synchronizing cash inflows and outflows is an important part of overall cash management. The contractor wants sufficient cash available to cover checks clearing the bank, but does not want an excess of cash sitting idle in the checking account. *Float* is defined as the difference between the balance shown in the contractor's books and the balance on the bank's books. It is created when the contractor has written checks that have not cleared the banking system, but have been deducted from the contractor's book balance for cash. This situation is evident at the end of each month when outstanding checks are present in the contractor's bank statement reconciliation.

Suppose the contractor writes, on the average, checks totalling $10,000 each day, and that it takes approximately five days for these checks to clear the banking system and be deducted from the contractor's bank account. After issuing five checks in one week, the contractor's own recorded cash balance will be $50,000 less than the bank's records. By the same token, if the contractor *receives* checks in the amount of $10,000 each day, but loses only three days while they are being deposited and cleared, the contractor's own books will reflect a balance that is $30,000 higher than the bank's balance. The net float created by the difference between the positive float of $50,000 and the negative float of $30,000 is a positive $20,000. This $20,000 float is a form of free financing. In essence, this means that the contractor's books could consistently show a negative cash balance of $10,000 while the bank is showing a positive balance of $10,000. This $20,000 amount, if it had to be financed through a bank loan costing 12% annually, would cost the contractor $2,400 per year in interest.

The contracting company's net float is a function of its ability to speed up collections on checks received and to slow down collections on checks written. This is why it is important to use a bank that makes deposited funds available as soon as possible. The extent to which the contractor can utilize float is sometimes determined by state banking laws. Appropriate laws should be reviewed to insure compliance.

Cash Conversion Cycle

The contractor's cash conversion cycle is the number of days during which the contractor must be able to finance ongoing internal operations. To calculate the conversion cycle, the contractor must determine the difference between the *incurrence of a cost* (e.g., the purchase of materials or labor) and the ultimate *collection of the billing* related to the incurred cost, minus the payment deferral period. For example, assume that Eastway Contractors typically has a 45-day period between the date it incurs a cost and the collection from the billing to the customer for that cost. This time frame might come about as follows: Eastway purchases materials for a job; it takes five days to incorporate the materials into the job, another 10 days to accomplish enough work for a progress billing, and 30 days to collect from the customer—thus, a 45-day (5 + 10 + 30) period. On the payment deferrals side, Eastway pays all invoices on the 30th day after the invoice date. In this case, Eastway's cash conversion cycle is 15 (45 − 30) days.

Figure 5.5 is a diagram of Eastway Contractors' cash conversion cycle. If Eastway is incurring costs of $5,000 per day, on average, the required cash investment in permanent working capital would be $75,000. This amount is calculated as the product of the cash conversion cycle of 15 days, times the average cost per day of $5,000.

It is one of the goals of effective cash management to reduce the duration of the cash conversion cycle and lower the required investment in working capital. Once again, this reduction is accomplished by accelerating collections and slowing disbursements.

Deceleration of Disbursements

The contractor should attempt to slow down disbursements to the point of payment on the last day possible, but *never past the date required to take advantage of cash discounts*. If the terms of purchase are 2/10, net 30, the contractor has a ten-day period within which to pay the invoice and deduct a 2% discount from the invoice amount, or 30 days to pay the net amount of the invoice. The cost of failing to take advantage of the 2% discount amounts to an annual effective cost of 36.72%. (This concept is explained in detail in Chapter 7.) If early payment discounts are not offered, the contractor should pay the invoice on the last day allowed by the invoice.

A number of other opportunities are available to the contractor for slowing down cash disbursements, including:

1. Use of a company credit card instead of providing employees with advances for travel or other expenses. Using the credit card generally provides the contractor with an additional 30 days' access to the funds involved.
2. Requesting more favorable credit terms from suppliers. A change of terms may be possible when a new supplier wants to establish a relationship, or when the level of the contractor's purchase increases significantly over previous levels and the past credit history has been favorable.
3. Timing the purchase of materials to coincide with the required need on the job site.

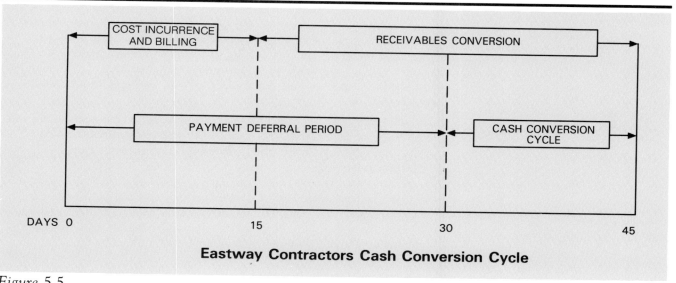

Eastway Contractors Cash Conversion Cycle

Figure 5.5

4. Including retention in subcontractors' agreements and specifying that the subcontractor is to be paid only after the contractor has received the appropriate progress payment.
5. Avoiding excessive investment in equipment and other capital assets.
6. Minimizing internal labor forces, which must be paid weekly and require various fringe benefits and insurances which must, in some cases, be prepaid.
7. Maximizing subcontractor forces which provide their own labor, material, and equipment on a credit basis. Subcontractors must complete a quantity of work during any given period, send an invoice, and wait to be paid. This clearly slows the disbursement of cash for direct costs and also eliminates a considerable amount of administrative expense.

Raising Cash

As discussed in the prior section, the contractor should take steps to estimate the firm's cash requirements and control receipts and disbursements to his greatest advantage. After these measures have been taken, it may be determined that there is a need for cash above the amount generated from ongoing operations. The first place to look for additional cash is *inside the company*. The contractor may be surprised at the resources that can be converted into cash without interrupting normal business operations. Close examination of cost control measures and better asset management are important steps in freeing up locked away cash.

After taking the normal steps to cut back on unnecessary expenditures, cost control may continue with the negotiation of early payment discounts with suppliers. Early payments may have the effect of increasing cash requirements in the very short run, but this effect should be more than offset by the long term benefits in cost reduction. Another approach is to put into practice more efficient purchasing practices in order to take advantage of quantity discounts on bulk purchases. These kinds of cost savings must be measured against the cost of purchasing and storing materials that are not currently required on jobs. The carrying cost of the idle materials may offset any gains received through the quantity purchase discount.

Effective asset management is discussed in detail in Chapter 6, but some important points are worth making as they relate to cash. Every asset used by the contractor represents a commitment of cash. If the asset is marginally productive, the use of the cash has been marginally productive. The contractor should examine the firm's assets to determine whether they could be used more effectively, or perhaps liquidated. Areas to analyze include:

1. *Equipment that is idle a vast majority of the time.* The contractor may be able to generate cash by selling the idle equipment and renting as the need arises. (Equipment can often be procured through an arrangement with subcontractors.)
2. *Accumulated materials and inventory that exceed a reasonable supply.* If possible, the inventory of excess materials should be used before buying new materials. When excess materials do accumulate, the contractor should attempt to use them on jobs, as appropriate, before they become damaged or otherwise unusable. Periodic advertising for the sale of miscellaneous materials may be a reasonable means of disposal.

If cash needs persist and cannot be met by internal sources, the contractor will have to arrange financing externally. Chapter 7, "Debt Management," addresses this subject in detail.

Short Term Investments

The term *surplus cash* refers to the cash balance that exceeds the sum needed to maintain ongoing operations. A well managed firm will invest its surplus cash in order to earn a return rather than leaving the unused sum idle in a checking account. In order to maximize the profitable use of surplus cash, the contractor needs to know the amount of surplus cash available and the length of time the amount will be available. This information is provided through the cash budgeting process.

The following discussion focuses on the investment of short term surplus cash (available for approximately one year or less), since this is the primary investment decision facing the contractor. Many types of temporary investment vehicles are available for the contractor. Figure 5.6 is a summary of several types of short term investment instruments, each of which is briefly described in the following section.

- *Money market funds* usually invest in a combination of U.S. Treasury and other government obligations, certificates of deposit, and commercial paper. They are available in denominations as low as $1,000, an attractive feature for smaller companies. These funds are managed by investment professionals who purchase in large quantities and sell small pieces to multiple investors. Although a service charge is usually involved, the yield is still higher than that of most U.S. government obligations. Some funds allow the investor to write checks against the fund such that their investment continues to earn interest until the check is presented for payment.

- *Certificates of deposit (CD's)* are negotiable interest-bearing obligations offered by commercial banks. Most certificates are issued by large commercial banks and the risk is low. However, since some risk does exist, the interest rate paid is usually higher than that paid on U.S. Treasury obligations. Although CD's usually mature within one year, the denominations vary and are often larger than practical for small companies.

- *U.S. Treasury Bills* are the most secure temporary investment. They can be purchased in denominations of $10,000 and can be readily traded on the open market. Treasury bills mature in three, six, nine, or twelve months, with slightly higher interest rates paid on the longer terms.

- *Commercial paper* is debt issued by (generally) large corporations for short periods of time. Because commercial paper is unsecured, it can be marketed by only the most successful corporations. Although it is considered relatively safe, the interest rate is higher than U.S. Treasury obligations. Denominations may be as low as $25,000, but are usually $100,000 or greater.

Since these short term investments are temporary "holding places" for the contractor's cash, the primary consideration is how quickly they can be converted to cash without loss of value. This criteria eliminates most long term securities from consideration because their maturities are normally too extensive and their values tend to vary widely. The three most important characteristics to consider in short term investing are *risk*, *marketability*, and the *term to maturity*.

Risk

A number of different types of risks pertain to the investment in short term securities. Default risk, interest rate risk, and liquidity risk are the most critical.

Default Risk: Default risk stems from the possibility that the issuer may be unable to pay interest or principal on time or at all. All securities except those issued by the U.S. Government are subject to some degree of default risk. The risk of default is one of the factors that influence the bond rating services, such as *Standard & Poor's Corporation* and *Moody's Investment Services*, when they assign a quality rating to securities.

Interest Rate Risk: This type of risk is associated with the fluctuation in market interest rates. For example, if the contractor invests $10,000 in a U.S. Treasury Bill with a maturity of one year and a yield of 8%, the investor's return is fixed regardless of changes in the market interest rates. If interest rates in the market on newly issued U.S. Treasury Bills of similar duration climb to 10%, then the value of the investor's 8% U.S. Treasury Bill declines. Who wants to invest in an instrument yielding 8% when similar investments are available yielding 10%? Thus, if the investor needs to liquidate the investment prior to the maturity

Summary of Short Term Investment Alternatives		
Investment Vehicle	**Normal Maturity**	**Description**
Money market mutual funds	Instant liquidity	Generally low risk, moderate yield, interest earned daily, good marketability.
Negotiable certificates of Deposit (CDs)	30–360 days	Risk depends upon bank, insured by FDIC up to $100,000, marketability fair since early withdrawal penalty may apply.
U.S. Treasury Bills	91 days–1 year	No default risk, excellent marketability.
Commercial paper	Up to 270 days	Default risk highest among short-term instruments but still low. Poor marketability. Usually issued by strongest firms.

Figure 5.6

date, he will in all likelihood have to discount the selling price of the security. The risk of a decline in value due to changing market interest rates is the basis of interest rate risk. The longer the term of the security, the greater the interest rate risk.

Liquidity Risk: Liquidity or marketability risk refers to the ease with which the investment can be converted into cash without suffering a significant price concession in the sale. Since the contractor is investing temporarily idle funds, and the business world is not a static environment, the contractor may need to liquidate the investments very rapidly. If the investment cannot be sold easily without marking down the price considerably, then the investment is said to have a high degree of liquidity risk.

Depending on the amount and duration of the availability of surplus cash, the contractor should select the investment vehicle that provides the highest expected return relative to the amount and type of risk that must be accepted.

Investment Services Offered by Banks: The size of the contracting company and its average account balance may provide the opportunity to take advantage of some creative short term investment vehicles. Some banks offer a vehicle called a *sweep account*. The purpose of the sweep account is to keep the depositor's otherwise idle cash invested until it is needed to cover checks which have been presented to the bank for payment. In this way, the contractor's checking account, in essence, maintains a zero balance. Each day as checks are presented to the bank for payment, the bank transfers enough money from the interest-bearing investment account to cover the checks. The sweep account is generally subject to a service charge by the bank, but for larger accounts, the income earned on the invested funds more than offsets the cost.

It is important that the contractor make a strong effort to understand all of the services offered by the bank. It may be well worth a meeting to discuss the company's relationship with the bank, and the variety of services that the bank offers. Many banking institutions have developed discount brokerage services, trust departments, investment counseling, and other related services. The contractor can maximize the relationship by better understanding the bank's services.

Summary

Cash management is the process of maximizing cash and the potential earnings from cash. The principles of good cash management include:
1. Planning the firm's cash needs.
2. Accelerating the receipt of cash.
3. Slowing down the disbursement of cash.
4. Generating cash.
5. Investing surplus cash.

There are four primary reasons for holding cash:
1. Transactions.
2. Compensating balance requirements.
3. Precaution against unexpected cash demands.
4. Investment and speculation.

The starting point in the cash management process is to identify the firm's potential cash receipts and cash requirements. The completion of a cash budget is a valuable process not just in facilitating the cash management function, but also because it forces the contractor to take a focused look at how the business operates.

The most common cash management advice given to firms is to accelerate collections and to slow down disbursements. Accelerating collections can often be achieved by the following methods.

- Effectively managing the firm's billing routine.
- Creating and following a disciplined accounts receivable policy.
- Depositing receipts in a timely manner and selecting a bank that does not have an overly restrictive waiting period prior to making deposited funds available for use.

The second directive, slowing down the disbursement of cash, may be accomplished by the following methods.

- Paying the invoice on the last day permissible without losing the cash discount. When discounts are not offered, the invoice should be paid on the last day of the allowable payment period.
- Using company credit cards instead of providing up-front cash advances.

Synchronization of cash flows means that the cash inflows and cash outflows are matched as closely as possible to avoid the holding of idle funds. *Float* is defined as the difference between the contractor's book cash balance and the bank's cash balance for the firm. Float is created by checks written by the contractor that have not cleared the bank system.

One of the primary goals of cash management is to reduce the contractor's cash conversion cycle. This is the period during which the contractor must absorb the cash cost of doing business.

When investing surplus cash, the contractor has a number of alternative investment vehicles to consider. Each is subject to a variety of risks. The primary risks to consider are:

- Default risk
- Interest rate risk
- Liquidity risk

The contractor's relationship with the bank is an important one. The banking industry has expanded the services it offers over the past several years. A thorough investigation of the available services should result in a stronger and more profitable banking relationship.

Chapter Six

ASSET MANAGEMENT

Chapter Six

ASSET MANAGEMENT

Assets are the economic resources employed by a firm to carry out its day-to-day activities. The assets of our example contracting firm, Eastway Contractors, Inc., include such resources as cash, accounts receivable, inventory, and fixed assets. These assets, among others, are included on the company's balance sheet (Chapter 3, Figure 3.2). At least one very important asset—human resources—is not listed on the typical balance sheet. For financial accounting purposes, human resources are not typically assigned a value, and therefore, do not appear among the listed assets on the balance sheet. This chapter will focus on those assets typically shown on a contractor's balance sheet.

Management is the function of planning, directing, and controlling resources so that the firm's objectives are accomplished in a timely and cost effective manner. Most of us operate in an environment with limited resources. Whether cash, facilities and equipment, or raw materials, there are limits to the available assets that can be employed to accomplish a particular objective. An effective manager uses the available assets most efficiently to meet the objectives of the organization. Management efficiency is achieved by:

1. Planning the acquisition of assets.
2. Directing the use of these assets in a cost effective manner.
3. Monitoring the rate of asset use by comparison with expectations.

We will begin with current asset management, i.e., the management of assets that will be converted into cash or used up within one year. The discussion will include the following topics.

- General Working Capital Policy
- Inventory Management
- Accounts Receivable Management

Cash is addressed as a separate topic in Chapter 5.

Current Asset Management

General Working Capital Policy

Working capital is calculated as the excess of current assets over current liabilities. It is the assets not required to satisfy current liabilities. This amount becomes a cushion for the expansion of operations and for flexibility; thus the term *working capital*.

A conservative working capital policy calls for the maintenance of a relatively large excess of current assets compared to current liabilities. An aggressive working capital policy calls for a relatively small excess. For example, under a conservative working capital policy, the contractor might carry one month's worth of materials in inventory. An aggressive policy might mean carrying two weeks worth of materials in inventory. The conservative contractor has the benefit of reduced risk from material shortages or time delays in materials. The aggressive contractor has an increased risk of stock shortage and time delay, but has a smaller commitment of funds tied up in idle assets. With fewer funds committed to idle assets, the overall return on investment increases, assuming that there are no negative consequences from the aggressive posture. The "best position" is debatable. The important point is that there is a trade-off of risk and return between alternate approaches.

Inventory Management

The term "inventory" generally brings to mind the local retail store with aisles and aisles of merchandise available for sale to the public. Inventory for the contractor may go beyond this basic concept of finished goods available for resale. The contractor's inventory most often relates only to the supply of miscellaneous unused building materials held for future use. However, if the contractor is in the business of building projects for speculation (as a developer), the accumulated cost of construction is transferred from the *Work-in-Process* account to an inventory account entitled *Completed Projects*. When the completed project is sold, the cost is transferred to the *Project Cost* account.

Inventory accounts are assets of the contractor. Therefore, the costs assigned to these accounts are not treated as expenses until revenue is generated from the project. For most of the contractor's work, projects constructed under a contract, the revenue is recognized as the work progresses, thereby preventing a build-up of inventory. Figure 6.1 shows the flow of these costs for the contractor.

The following section highlights the fundamentals of a contractor's cost accounting system as it relates to the accumulation of costs for inventory purposes. This section focuses primarily on construction for speculation. Accounting systems are discussed in more detail in Chapter 2.

Since the contractor is in the business of combining materials, labor, and operating costs to construct a project, it is important for him to maintain a cost accounting system that will answer the following key inventory questions.

1. To date, what is the total cost (material, labor, subcontractors, operating expenses) incurred on a specific project?
2. Based on original cost projections, what percentage complete is the project in terms of costs to date actually incurred?
3. What is the fully burdened cost of jobs completed thus far?

The answers to these questions are important to the contractor from the standpoint of both operations management and project financing. A cost accounting system that provides this information requires a Job Cost Ledger with cost detail on each job. This ledger should be prepared and reconciled with the financial reports, on a monthly basis. Not only does the contractor need to capture accurate and reliable cost data on each job, but the data must be prepared in a timely manner. It does very little good after the fact to know that the last five projects were sold for 10% below the fully burdened cost.

In order for information to be useful, it must be relevant. To be relevant, the information must (1) make a difference to the decision that needs to be made, and (2) be received in time to have an impact. A well developed cost accounting system will meet both of these objectives.

Types of Inventories

The contractor may require three different categories of inventory:

1. Raw Materials
2. Work-in-Process (WIP)
3. Completed Projects

The first category is the supply of *raw materials and supplies* maintained on hand. Depending on the form of operation and the size of the contracting firm, the raw materials inventory might consist only of the materials on the job site that are not yet part of the project; or it could include the supply of raw materials stored in a large warehouse. In either case, the raw materials inventory represents the materials for use on a job that have not yet become part of the job.

The second inventory classification for the contractor is *Work-In-Process (WIP)*. WIP represents the fully burdened cost of construction incurred towards the completion of a specific project. The term *fully burdened*

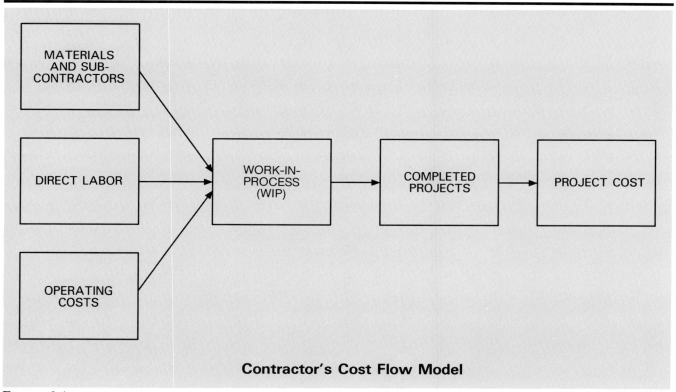

Contractor's Cost Flow Model

Figure 6.1

means that the WIP account valuation should include the following three cost classifications:

1. Materials and Subcontractors
2. Primary Labor
3. Operating Expenses (indirect construction costs and General and Administrative expenses)

When a speculative internal job is completed, the accumulated WIP cost for the job is transferred to the final inventory account, *Completed Projects*.

Current federal tax regulations require that inventory be costed using this "full absorption" approach. This approach has a direct impact on the tax liability of the firm, since the classification of the cost affects the timing of its deductibility for tax purposes. Once again, an inventory cost is not expensed until it is matched against revenue.

The use of the WIP Account varies depending on the contractor's method of operation. A contractor engaged in the speculative building of projects for resale clearly needs to maintain an inventory account for the work-in-process. A contractor working under a contract to build a specific project for a prospective owner may not maintain an inventory WIP Account, preferring instead to expense the construction costs incurred and recognize profit on the construction as work progresses. In both cases, there is clearly a need to maintain a detailed job cost accounting system.

Cost Classifications

The major cost classifications of materials and subcontractors, direct labor, and operating costs are either directly assigned or appropriately allocated to work-in-process. Direct materials, subcontractors, and direct labor should be specifically charged to a job when incurred. For the contractor with no central materials storage warehouse, this procedure involves recording a specific job number on the packing slip when materials are received from the supplier. Ideally, the contractor prepares a purchase invoice to document the nature and quantity of materials requested, the price, and the job to receive the materials. If possible, the supplier should note the job number on the billing.

Direct labor should be charged to jobs based on actual hours worked. Time cards should be completed by each employee with all direct labor charged to the appropriate job. Nondirect labor is considered part of the operating cost category. It is advisable to direct charge as much cost as practical.

Non-direct Costs: After the costs that can be directly charged to the job are tracked, non-direct charges must be allocated. This allocation process is accomplished in the sequence outlined below. Chapter 4 contains a detailed discussion of operating cost allocation.

1. Establish a predetermined allocation.
2. Based on the activity for the period, calculate the operating cost to be allocated to a specific job within the *Work-in-Process* account.
3. Complete the job cost ledger noting the month's operating cost allocation.

Accounts Receivable Management

Accounts Receivable are created when goods or services are extended to a customer on credit. Although some contractors may not realize it, they routinely extend credit to their customers. This is because construction contracts typically require the firm to perform a certain amount of work *before* billing for it. After the invoice is sent, the customer is normally granted several weeks before payment is required. During this time, the contractor is, in essence, financing a portion of the project for the customer. However, if the contractor stages his receivables and payables properly, he will not be making the bulk of the creditor payments associated with any given draw until payment has been received from the owner.

The extension of credit is a key element in a firm's operations; thus, a sound credit policy must be established. The policy should cover these areas.

1. Credit Standards
2. Credit Terms
3. Credit Account Tracking
4. Collection
5. Timely Billing Practice

Credit Standards: Credit Standards are the basic credentials that a customer must possess in order to receive credit. To receive credit, a customer should provide documentation acceptable to the contractor. The contractor must follow up by verifying the information and contacting the applicant's references. The decision to extend credit and the amount of credit to extend should be based on the evaluation of the documentation and reference information. Most importantly, a general contractor should require verification of secure financing from the owner's lending institution. This topic is covered later in this chapter in the section on *Contractual Protection*.

Credit information about prospective customers can generally be obtained from several of the following sources:

- Dun and Bradstreet
- Credit Bureaus
- Banks
- Other Contractors
- Suppliers
- Financial Statements

Once credit is established, the contractor should follow up periodically, requesting updated financial information from credit customers at least once each year. A specific individual within the firm should be charged with regularly reviewing the credit status of existing customers. Evaluation is generally centered around the five C's of credit listed below.

1. **Character** refers to the likelihood that the customer will feel morally bound to honor the obligation. Credit reports generally highlight whether or not a potential customer has a track record of honoring his obligations. Character is often considered the most important of the five C's, but it is sometimes the most difficult to evaluate.
2. **Capacity** refers to the customer's ability to pay. This determination is often based on an objective judgment of the customer's overall business stability and the security of his financing arrangements.

3. **Capital** refers to the customer's financial strength. This strength might be measured by a combination of ratio analysis and an examination of the customer's existing debt exposure.
4. **Collateral** refers to the assets that the customer owns that could be used to secure the loan.
5. **Conditions** refer to the general state of the economy and the specific state of the industry. Favorable conditions usually indicate continued business prosperity, while unfavorable conditions indicate increased risk of a business downturn.

Credit Terms: Credit Terms specify the period for which credit is extended and the discount, if any, given for early payment. While it is uncommon for contractors to offer discount terms, this approach offers the advantage of accelerated cash collection. The disadvantage is that if margins are already tight, the discount will mean a further reduction in overall job profitability. It is, therefore, in the contractor's best interest not to offer discounts, and to extend credit on a net 10 basis.

It is important that credit terms be specifically communicated to the owner and his lender when the contract is initiated. Accounts Receivable is an unproductive asset. The funds that are tied up in Accounts Receivable could be put to better use if invested in interest-yielding securities or used to reduce liabilities. If the average balance in Accounts Receivable is $100,000 and the cost of capital to the firm is 10%, then the annual cost of carrying the receivables is $10,000.

A customer who extends payment beyond the specified term of the agreement should pay an interest charge. The terms should be established at the start, as it is difficult to add an interest or penalty charge after the contract is active. It is always advisable to pursue diligently the first payment on a project in order to firmly establish the system and to show determination in adhering to it.

Credit Account Tracking: Credit Account Tracking provides information on the current status of each customer's outstanding account balance. The tracking system should be maintained by one individual who is familiar with the contractual terms and receivables on all projects. One of the most important credit management reports is the *Schedule of Aged Accounts Receivable*. Figure 6.2 depicts a standard aging schedule. The purposes of the schedule include:

- Providing a listing of all accounts that make up the balance of Accounts Receivable as shown on the firm's balance sheet.
- Detailing the delinquency of any account balances beyond the normal payment period.
- Providing a visual report that highlights potential credit collection problems before they became excessive.

In Figure 6.2, it is assumed that 0 to 14 days is the normal collection period for contractors. If a company's terms are different from the 14-day period, the aging schedule can be modified to make it more meaningful. If customers fail to pay their accounts, the balance moves from left to right, providing a very visible indication of increasing delinquency. The contractor should have very specific responses to this movement of customer accounts beyond the 0 to 14-day aging, as explained in the next section on *Collection*.

Collection: Collection refers to the actions taken by the contractor in the event of nonpayment by the customer. Sometimes this action involves a lien against the job or legal action against the owner. In other cases, it means enlisting the services of a collection agency or lawyer. Hopefully, neither of these options will be necessary. The specific timing and extent of follow-up actions will depend on factors such as the terms of the contract and the dollar amount of the delinquency. It is important to remember that management needs the information provided by the aging schedule to make informed credit decisions.

The first step is to send a letter, informing the customer of the nonpayment. Often, a personal handwritten note from the president is more effective than a typed *past due* notice. If this letter, together with a personal phone call, does not have the desired effect, then a formal collection letter should be sent.

The collection letter should note the debt and the date of the obligation, and it should seek an explanation for the lack of payment. The letter must be firm, but non-threatening. There is no need to acknowledge the customer's financial problems if they are known to exist or to offer to accept a reduced payment to clear the account. The contractor may, however, have to be willing to accept an installment payment from the customer.

If there is no response to the formal collection letter, then a final collection letter should be sent, perhaps certified. The letter should set a date for final payment and state that payment in full is expected by that time. It is important to follow up, turning the account over to a collection agency or attorney if the letter states that this will be done. These options for collection are generally expensive (perhaps as much as 50% of the account balance or more) and should, therefore, be used with discretion.

When an owner fails to make payment, this information should be quickly communicated to all affected parties within the firm. The president, project manager, and supervisor should all be informed, and subsequent service and actions should reflect the nonpayment. Work should be stopped as soon as contractually allowed if a serious payment problem appears to have developed.

Schedule of Aged Accounts Receivable					
Customer	Total Due	0–14 Days	14–30 Days	30–60 Days	Over 60 Days
Owner # 1	10,000	8,000	2,000		
Owner # 2	80,000	65,000	10,000	5,000	
Owner # 3	115,000	100,000	15,000		
Total	205,000	173,000	27,000	5,000	

Figure 6.2

There are often legitimate or unexpected reasons why payments are delayed. Efforts should be made to anticipate and avoid these situations whenever possible. Examples of such circumstances are:

1. Defective work which has not been corrected.
2. Claims filed by third parties.
3. Failure of the contractor to pay subcontractors or suppliers.
4. Accidents that injure workers or cause damage to the project, and that may result in litigation.
5. Evidence that the work cannot be finished in time and that remaining funds will not cover liquidated damages.
6. Persistent failure to perform in accordance with the contract.
7. Disputes over quality, value of work completed, or errors in billing.
8. Failure to provide necessary documentation, e.g., certifications, affidavits, and approvals.

Timely Billing Practices: Minimizing the cost of carrying a firm's accounts receivable entails more than having an effective credit policy. Timely billing practices are an important part of the overall management effort. Many contracts are based on progress payments to be made when specific "milestones" are accomplished. AIA documents G702 and G703 are commonly used for billing draws in the commercial construction industry. The contractor must establish effective communication between the project manager and the billing department to ensure that each job is billed as soon as possible. This "time is of the essence" philosophy is very apparent when cash flow problems exist, but tends to get less attention when cash flow is adequate. For the effective manager, there should be no difference. The policy should be to bill *all* work as soon as allowable and maximize the cash flow to the firm. This means that the contractor should not always wait until the end of the month to do the billing for the month. It is doubtful that the efficiencies gained by doing all the billing at one time will offset the benefits of the increased cash flow generated by accelerating the billing process. Furthermore, it is easier for the customer to relate the billing to the work performed when the billing occurs soon after the work is completed.

Contractors should be aggressive in the amount of their billings. Overbilling, or front end loading, can be used to finance jobs whenever possible. In the end, only the full contract price can be billed, but it is to the contractor's advantage to accelerate the contract receipts.

To minimize problems with billing and collection, the contractor should adhere to the following guidelines:

- Plan the anticipated draw schedule and communicate it to the owner and lender.
- Be familiar with the client's invoice approval process so that all key players are identified.
- Make sure that the client and his lender fully understand the payment terms and conditions.
- Make sure that all billings are accurate and in accordance with established procedures.
- Be quick to respond to any delay in payment.
- Be particularly careful to "walk through" all steps of the first invoice to insure timely payment.
- Make sure that accounts receivable management is a high priority for a particular, assigned individual.

Current assets provide the firm's day-to-day operating capital. Inefficient management of current assets, including accounts receivable, leads to unnecessary cash flow problems and possible interruptions in operations. Such inefficiencies lead to less than optimal financing arrangements and, ultimately, a lower return on investment for the owners.

Contractual Protection

There may be no aspect of accounts receivable control more important to the contractor than the contractual terms under which payments are to be made by the owner. All too often, contractors make unjustified assumptions about their customers' credit-worthiness and business principles. The fact is that many firms are forced into bankruptcy by owners who become far too leveraged or are unscrupulous in their financial dealings. Fortunately, proper contractual language can eliminate most of these risks.

A number of standard contractual agreements have been prepared by the American Institute of Architects. These forms are commonly used for most public projects and for traditional, competitively bid private projects. In negotiated design/build or construction management relationships, however, the contractor may have far more input into the content and form of the agreement. Regardless of what form is to be used, the contractor should ensure that certain important points affecting the security of payments are included. Some of these points are listed below.

- Amount and method of payment.
- Type of requisition form.
- Schedule of values.
- Timing of draw submissions and payments.
- Retention terms—amount and time of release.
- Performance bonds.
- Payment for material stored on site.
- Payment of non-disputed portions of draw.
- Recourse in the event of default.
- Arbitration in the event of dispute.
- Final inspection and punch list procedures.
- Payment for engineered materials or equipment.
- Insurance coverage for materials stored on site.

Retainage is a particularly troublesome issue for contractors, and must be carefully negotiated prior to execution of any contract. The 10% standard retention on each draw far exceeds the 3% to 5% net profit earned by most contractors, causing a slow, but sure, drain on working capital. Many managers have been fortunate, however, in negotiating a number of more favorable retention terms, such as:

- The elimination of any retention.
- Submission of an affidavit each month or upon substantial completion stating that all bills have been paid.
- Retainage on labor only, not material.
- Placement of retainage in an interest-bearing escrow account, with the interest accruing to the contractor.
- Reduction from 10% to 5% throughout the project, or 10% on only the first 50% of the project, and nothing thereafter.
- Retainage of a specified amount in lieu of a percentage.
- Release of portions of the retainage upon achievement of specified milestones.
- Provisions of a letter of credit in lieu of retainage.

Appendix C is a sample fixed cost contract showing a number of clauses that protect the contractor. Many of the issues covered in this discussion are included in the sample contract. Although this form may not be wholly acceptable to every owner, it is advisable that as much of the intent as possible be retained, even if it must be added in the form of amendments to other standard contractual forms.

Capital Asset Management

Capital assets are those that will provide service to the construction firm for years into the future. The distinction separating current from noncurrent (capital) assets is that current assets are consumed or turned over within the year, while noncurrent assets provide service beyond the one-year period.

Since the service period for capital assets is so long, the decision to acquire a noncurrent asset carries a more prolonged risk element than that associated with the acquisition of a current asset. *Risk* generally refers to the probability of the *actual* outcome of an event being different than the *expected* outcome of the event. The greater the uncertainty, the greater the risk. Most of us feel more comfortable forecasting business conditions over the next 6 to 12 months than we would the next 12 to 24 months. By their very nature, capital assets acquired today will be in service during these future periods. Thus, the decisions made today have a significant impact on the contractor's positioning and risk in future business operations.

The following discussion of capital asset management is divided into two sections. The first section focuses strictly on the economics of capital asset acquisition, while the second deals with the more general managerial procedures concerning capital assets.

Financial Analysis of Capital Assets

Typical capital asset acquisition decisions for the contractor include:
- **Equipment expansion and replacement decisions.** Should old equipment be replaced or expanded upon now or in the future?
- **Plant expansion decisions.** Should the building and other plant facilities be expanded, and will the benefits gained by the expansion offset the cost?
- **Purchase vs. lease decisions.** Is it more cost effective in the long run to lease or purchase heavy equipment?

Clearly, acquiring capital assets involves the commitment of funds in the current period, with returns being generated largely in future periods. The problem becomes one of comparing the *costs* of a capital asset with the *benefits* to be generated by the asset. The most immediate problem is that there is a difference in the timing of the cash outflows and the cash inflows. No discussion of decisions related to capital asset acquisition, referred to as *capital budgeting*, can begin without a thorough understanding of the time value of money.

Time Value of Money

History shows that it is very rare when inflation is not a factor in our economy. A dollar received today is, therefore, worth more than a dollar received in the future. This is true for three reasons. First, the purchasing power of a dollar deteriorates over time. Consequently, more dollars are needed to purchase the same quantity of product in the future. Second, the future holds uncertainty. The further one plans into the future, the greater the uncertainty. Therefore, as the benefits to be derived from an investment in a capital asset are received further in the future, they become more uncertain. Third, a dollar received today can be invested and become an income producing liquid asset for the firm.

Two basic concepts are important to the discussion of the time value of money. These are *future value* and *present value*. Future value represents the future earnings of the dollar invested today. For example, if $100 is invested today in an interest-bearing account paying 5% annually, the account will grow to the following future values at the end of each of the next three years:

Original investment	**$100.00**
Interest for the first year:	
$100.00 × 0.05	5.00
Balance at the end of year one	**$105.00**
Interest for the second year:	
$105.00 × 0.05	5.25
Balance at the end of year two	**$110.25**
Interest for the third year:	
$110.25 × 0.05	5.51
Balance at the end of year three	**$115.76**

The interest for each year is greater than the interest for the previous year. This increase is due to the compounding of the interest from the previous year. *Compounding* means the earning of current period interest on interest that has already been earned in past periods. The more often interest is compounded, the faster the money grows.

It would be cumbersome to continue this procedure to discover the value at the end of the tenth year. Fortunately, tables have been established to simplify this calculation. Figure 6.3 shows the factors used to calculate the future value of an amount invested today. The table shows the interest rates in columns, and time periods in rows. For our previous example, the applicable factor would be based on an interest rate of 5% and three periods. Since, in our example, the interest is compounded annually, the periods would be equal to the number of years, times one. If our interest is compounded semi-annually, each year would represent two periods; quarterly would be four periods, etc.

In Figure 6.3, the factor represented by 5% and three periods is 1.1576. To use this factor we employ the formula:

Future Value = Present Value × Future Value Interest Factor

From this point, the following abbreviations are used:

Future Value	=	FV
Present Value	=	PV
Future Value Interest Factor	=	FVIF

Filling in the formula using $100 for PV, we have

FV $100 (5%, 3 periods) = $100 x FVIF (5%, 3 periods)
= $100 x 1.1576
= $115.76

Multiplying $100 by the factor 1.1576, we find the future value is $115.76, the same answer as originally calculated.

The time value of money is further illustrated in the following examples, using our hypothetical contracting company, Eastway Contractors.

Example 1

Eastway contractors has recently completed a successful project that generated a cash surplus of $20,000. This surplus is now available for investment. Eastway realizes that it may be necessary to expand plant facilities in the next two years and is, therefore, interested in accumulating cash to help fund the expansion. How much cash will Eastway have available at the end of two years if the $20,000 is invested at 8%, compounded annually, for the entire two-year period?

Solution:

$$\text{FV } \$20{,}000 \; (8\%, \; 2 \text{ periods}) = \$20{,}000 \times \text{FVIF } (8\%, \; 2 \text{ periods})$$
$$= \$20{,}000 \times 1.1664$$
$$= \$23{,}328$$

According to this computation, Eastway will have $23,328 available at the end of the second year.

Future Value of $1

Periods	2%	3%	4%	5%	6%	7%	8%	10%	12%
1	1.0200	1.0300	1.0400	1.0500	1.0600	1.0700	1.0800	1.1000	1.1200
2	1.0404	1.0609	1.0816	1.1025	1.1236	1.1449	1.1664	1.2100	1.2544
3	1.0612	1.0927	1.1249	1.1576	1.1910	1.2250	1.2597	1.3310	1.4049
4	1.0824	1.1255	1.1699	1.2155	1.2625	1.3108	1.3605	1.4641	1.5735
5	1.1041	1.1593	1.2167	1.2763	1.3382	1.4026	1.4693	1.6105	1.7623
6	1.1262	1.1941	1.2653	1.3401	1.4185	1.5007	1.5869	1.7716	1.9738
7	1.1487	1.2299	1.3159	1.4071	1.5036	1.6058	1.7138	1.9487	2.2107
8	1.1717	1.2668	1.3686	1.4775	1.5938	1.7182	1.8509	2.1436	2.4760
9	1.1951	1.3048	1.4233	1.5513	1.6895	1.8385	1.9990	2.3579	2.7731
10	1.2190	1.3439	1.4802	1.6289	1.7908	1.9672	2.1589	2.5937	3.1058
11	1.2434	1.3842	1.5395	1.7103	1.8983	2.1049	2.3316	2.8531	3.4785
12	1.2682	1.4258	1.6010	1.7959	2.0122	2.2522	2.5182	3.1384	3.8960
13	1.2936	1.4685	1.6651	1.8856	2.1329	2.4098	2.7196	3.4523	4.3635
14	1.3195	1.5126	1.7317	1.9799	2.2609	2.5785	2.9372	3.7975	4.8871
15	1.3459	1.5580	1.8009	2.0789	2.3966	2.7590	3.1722	4.1772	5.4736
16	1.3728	1.6047	1.8730	2.1829	2.5404	2.9522	3.4259	4.5950	6.1304
17	1.4002	1.6528	1.9479	2.2920	2.6928	3.1588	3.7000	5.0545	6.8660
18	1.4282	1.7024	2.0258	2.4066	2.8543	3.3799	3.9960	5.5599	7.6900
19	1.4568	1.7535	2.1068	2.5270	3.0256	3.6165	4.3157	6.1159	8.6128
20	1.4859	1.8061	2.1911	2.6533	3.2071	3.8697	4.6610	6.7275	9.6463
25	1.6406	2.0938	2.6658	3.3864	4.2919	5.4274	6.8485	10.835	17.000
30	1.8114	2.4273	3.2434	4.3219	5.7435	7.6123	10.063	17.449	29.960
40	2.2080	3.2620	4.8010	7.0400	10.286	14.974	21.725	45.259	93.051
50	2.6916	4.3839	7.1067	11.467	18.420	29.457	46.902	117.39	289.00
60	3.2810	5.8916	10.520	18.679	32.988	57.946	101.26	304.48	897.60

Figure 6.3

If Eastway chooses a financial institution that compounds the investment semi-annually, there would be four compounding periods over the two years. Figure 6.3 can again be used for this calculation, but the number of periods must be doubled and the interest rate cut in half, to 4%. (The interest rate must be divided by two because the compounding is taking place every six months instead of once each year.)

The new solution for semi-compounding would be:

$$\text{FV } \$20{,}000 \text{ (4\%, 4 periods)} = \$20{,}000 \times \text{FVIF (4\%, 4 periods)}$$
$$\text{for semi-} \qquad\qquad\qquad = \$20{,}000 \times 1.1699$$
$$\text{compounding} \qquad\qquad = \$23{,}398$$

The use of semi-annual compounding results in an additional $70 of interest income over the two years.

Annuities: In the prior examples, the investment is a one-time amount. Sometimes, an investment is made in a series of consecutive periods. The term *annuity* is used to describe a series of investments, or cash flows. To illustrate, assume that Eastway Contractors invested $5,000 in an interest-bearing account each year for the next ten years earning an average rate of return of 8%. To determine the value of the account at the end of the ten-year period involves the use of the future value of an annuity calculation. This calculation is similar to the *future value* calculation used previously and is stated as follows:

Future Value of an Annuity =
Annuity × Future Value Interest Factor for an Annuity

The abbreviations for the variables are as follows:

Future Value of an Annuity $\qquad\qquad$ = FVA
Annuity $\qquad\qquad\qquad\qquad\qquad\qquad$ = A
Future Value Interest Factor for the Annuity = FVIFA

Figure 6.4 provides the factors used in Future Value of an Annuity calculations.

The calculation used to determine the value of Eastway's account after depositing $5,000 per year for 5 years is as follows.

$$\text{FVA } \$5{,}000 \text{ (8\%, 5 periods)} = \$5{,}000 \times \text{FVIFA (8\%, 5 periods)}$$
$$= \$5{,}000 \times 5.8666$$
$$= \$29{,}333$$

The factor, 5.8666, is based on the interest rate of 8% and the number of annual payment periods, five.

The account balance after the fifth deposit of $5,000 is $29,333. After ten deposits of $5,000, the account balance is $72,435 ($5,000 × 14.487).

Example 2

Eastway Contractors has entered into a financing arrangement with a lending institution which agrees to provide $250,000 in debt financing to the firm. Repayment of the principal amount will occur at the end of five years with interest-only payments being made each quarter until the principal payoff. Eastway wants to provide for the payoff by depositing an equal amount into a bank account each year for the next five years so that after the final deposit, the account balance is equal to the required $250,000. Deposits into the account will begin one year from today and are expected to earn an average return of 8% over the five years.

In this example, the future value of the annuity is known. The unknown is the amount of the annual deposit necessary to fund the future payoff of the bank note's principal balance. This unknown amount can be calculated using the Future Value of an Annuity formula.

Two of the three variables within the formula are known. The FVA is equal to $250,000 and the FVIFA is determined to be 5.8666 based on an 8% interest rate and five periods. By plugging these known variables into the formula, the solution is:

$$FVA = PMT \times FVIFA$$
$$\$250,000 = A \times 5.8666$$
$$A = \frac{\$250,000}{5.8666}$$
$$A = \$42,614$$

Future Value of an Annuity of $1

Periods	2%	3%	4%	5%	6%	7%	8%	10%	12%
1	1.0000	1.0000	1.0000	1.0000	1.0000	1.0000	1.0000	1.0000	1.0000
2	2.0200	2.0300	2.0400	2.0500	2.0600	2.0700	2.0800	2.1000	2.1200
3	3.0604	3.0909	3.1216	3.1525	3.1836	3.2149	3.2464	3.3100	3.3744
4	4.1216	4.1836	4.2465	4.3101	4.3746	4.4399	4.5061	4.6410	4.7793
5	5.2040	5.3091	5.4163	5.5256	5.6371	5.7507	5.8666	6.1051	6.3528
6	6.3081	6.4684	6.6330	6.8019	6.9753	7.1533	7.3359	7.7156	8.1152
7	7.4343	7.6625	7.8983	8.1420	8.3938	8.6540	8.9228	9.4872	10.089
8	8.5830	8.8923	9.2142	9.5491	9.8975	10.260	10.637	11.436	12.300
9	9.7546	10.159	10.583	11.027	11.491	11.978	12.488	13.579	14.776
10	10.950	11.464	12.006	12.578	13.181	13.816	14.487	15.937	17.549
11	12.169	12.808	13.486	14.207	14.972	15.784	16.645	18.531	20.655
12	13.412	14.192	15.026	15.917	16.870	17.888	18.977	21.384	24.133
13	14.680	15.618	16.627	17.713	18.882	20.141	21.495	24.523	28.029
14	15.974	17.086	18.292	19.599	21.015	22.550	24.215	27.975	32.393
15	17.293	18.599	20.024	21.579	23.276	25.129	27.152	31.772	37.280
16	18.639	20.157	21.825	23.657	25.673	27.888	30.324	35.950	42.753
17	20.012	21.762	23.698	25.840	28.213	30.840	33.750	40.545	48.884
18	21.412	23.414	25.645	28.132	30.906	33.999	37.450	45.599	55.750
19	22.841	25.117	27.671	30.539	33.760	37.379	41.446	51.159	63.440
20	24.297	26.870	29.778	33.066	36.786	40.995	45.762	57.275	72.052
25	32.030	36.459	41.646	47.727	54.865	63.249	73.106	98.347	133.33
30	40.568	47.575	56.085	66.439	79.058	94.461	113.28	164.49	241.33
40	60.402	75.401	95.026	120.80	154.76	199.64	259.06	442.59	767.09
50	84.579	112.80	152.67	209.35	290.34	406.53	573.77	1163.9	2400.0
60	114.05	163.05	237.99	353.58	533.13	813.52	1253.2	3034.8	7471.6

Figure 6.4

Solving for A involves isolating it from the other variables. This is accomplished by dividing each side of the equation by 5.8666.

While calculating the future value of a single investment or a series of investments is useful for the contractor, the primary focus in capital budgeting is on the *present value of future cash flows*. It is the present value calculations that enable a contractor to evaluate the future benefits of an investment against its current cost.

Capital Budgeting

Capital budgeting involves committing funds in the current period with the expectation of receiving a desired return on those funds in the future. For the contractor, this means investing funds in business assets that are expected to provide a positive contribution to overall business operations and profitability over a period of years.

In most cases, the contractor has a limited budget within which to operate and is, therefore, eager to commit funds to the assets that promise the highest expected rate of return. Generally, the measure of expected return is, at best, only an indicator of the *relative merit* of the project, since the capital budgeting process tends to be subjective. This subjectivity is due to the following variables.

1. Forecasting the firm's cost of capital in the future.
2. Forecasting the amount and timing of a project's cash inflows and outflows.
3. Anticipating the possible technical obsolescence of the acquired equipment, as well as its useful life.
4. Potential changes in the firm's corporate tax rate.

The fact that it is a subjective process does not invalidate the usefulness of capital budgeting. On the contrary, requiring the manager's involvement ensures that steps are taken to effectively plan the acquisition and use of capital assets. This budgeting exercise creates a benchmark against which actual asset performance can be measured. In this way, the relative effectiveness of the assets employed by the firm can be measured.

A number of methods are used to evaluate alternatives for investing in capital assets. The most useful methods are those employing *discounted cash flow*. Since the contractor is weighing today's investment against future benefits, it is necessary to compare dollars in terms of purchasing power. For example, let us assume that Eastway Contractors has an opportunity to invest $50,000 in either of two mutually exclusive pieces of heavy equipment, such as a track loader or a backhoe. *Mutually exclusive* means that the pieces of equipment will provide similar benefits and only one or the other will be acquired. Each asset will provide labor cost savings over a five-year period. At the end of three years, the asset will have no salvage value. When evaluating investments of this type, Eastway Contractors uses a discount rate of 12%. This percentage is decided based on (a) a rate of return commensurate with the risk of the investment and (b) a rate that will cover the firm's cost of capital, which is the rate that the contractor would have to pay when financing the asset.

Projecting the net cash flows associated with each investment alternative is the first step in evaluating the asset acquisitions. The projected net cash flows for the track loader and backhoe, due largely to labor and rental savings, are as follows:

		Track Loader	Backhoe
Cash inflow	(period 1)	20,000	14,200
Cash inflow	(period 2)	18,000	14,200
Cash inflow	(period 3)	14,000	14,200
Cash inflow	(period 4)	10,000	14,200
Cash inflow	(period 5)	8,000	14,200
Total net cash inflows		**70,000**	**71,000**

The backhoe produces total net cash inflows $1,000 greater than the net cash inflows produced by the track loader. Does this mean that the backhoe should be accepted over the track loader? The answer is not readily apparent since the timing of the cash inflows is different for each of these investments.

To convert future cash flows to present dollars, the following equation is used:

Present Value = Future Value X Present Value Interest Factor.

Abbreviations for the formula variables are as follows:

Present Value = PV
Future Value = FV
Present Value Interest Factor = PVIF

Figure 6.5 provides the factors used to calculate the present value of a single sum to be received in the future. The structure of the Present Value table is the same as the Future Value tables. It should be noted that no factor on the table is greater than .9901. This is due to the fact that unless the assumed discount factor is 0%, no future dollar will be worth as much as a present day dollar. In order to properly evaluate the investment in the track loader or backhoe, the future cash inflows must be discounted back to their present value. The schedule in Figure 6.6 shows the Present Value calculations for the future cash flows generated by both investments.

For both investments, the present value of the net cash inflows are greater than the initial outlay of $50,000. This tells Eastway Contractors that both investments generate a return greater than the 12% required rate. Since the present value of the cash inflows is greater for the track loader than the backhoe, Eastway Contractors should select the track loader, all other factors (delivery time of equipment, maintenance costs, etc.) being equal. Prior to the present value analysis, the backhoe appears to yield the greatest return. However, when the time value of money is considered, the track loader is clearly the better investment.

Present Value of $1

Periods	1%	2%	3%	4%	5%	6%	7%	8%	10%	12%
1	.99010	.98039	.97087	.96154	.95238	.94340	.93458	.92593	.90909	.89286
2	.98030	.96117	.94260	.92456	.90703	.89000	.87344	.85734	.82645	.79719
3	.97059	.94232	.91514	.88900	.86384	.83962	.81630	.79383	.75131	.71178
4	.96098	.92385	.88849	.85480	.82270	.79209	.76290	.73503	.68301	.63552
5	.95147	.90573	.86261	.82193	.78353	.74726	.71299	.68058	.62092	.56743
6	.94205	.88797	.83748	.79031	.74622	.70496	.66634	.63017	.56447	.50663
7	.93272	.87056	.81309	.75992	.71068	.66506	.62275	.58349	.51316	.45235
8	.92348	.85349	.78941	.73069	.67684	.62741	.58201	.54027	.46651	.40388
9	.91434	.83676	.76642	.70259	.64461	.59190	.54393	.50025	.42410	.36061
10	.90529	.82035	.74409	.67556	.61391	.55839	.50835	.46319	.38554	.32197
11	.89632	.80426	.72242	.64958	.58468	.52679	.47509	.42888	.35049	.28748
12	.88745	.78849	.70138	.62460	.55684	.49697	.44401	.39711	.31863	.25668
13	.87866	.77303	.68095	.60057	.53032	.46884	.41496	.36770	.28966	.22917
14	.86996	.75788	.66112	.57748	.50507	.44230	.38782	.34046	.26333	.20462
15	.86135	.74301	.64186	.55526	.48102	.41727	.36245	.31524	.23939	.18270
16	.85282	.72845	.62317	.53391	.45811	.39365	.33873	.29189	.21763	.16312
17	.84438	.71416	.60502	.51337	.43630	.37136	.31657	.27027	.19784	.14564
18	.83602	.70016	.58739	.49363	.41552	.35034	.29586	.25025	.17986	.13004
19	.82774	.68643	.57029	.47464	.39573	.33051	.27651	.23171	.16351	.11611
20	.81954	.67297	.55368	.45639	.37689	.31180	.25842	.21455	.14864	.10367
25	.77977	.60953	.47761	.37512	.29530	.23300	.18425	.14602	.09230	.05882
30	.74192	.55207	.41199	.30832	.23138	.17411	.13137	.09938	.05731	.03338
40	.67165	.45289	.30656	.20829	.14205	.09722	.06678	.04603	.02209	.01075
50	.60804	.37153	.22811	.14071	.08720	.05429	.03395	.02132	.00852	.00346
60	.55045	.30478	.16973	.09506	.05354	.03031	.01726	.00988	.00328	.00111

Figure 6.5

Investment A

Track loader

Period	Cash Inflow	Present Value Factor	Present Value
1	$20,000	.8929	$17,858
2	18,000	.7972	14,358
3	14,000	.7118	9,965
4	10,000	.6355	6,355
5	8,000	.5674	4,539
	70,000		Total 53,067

Investment B

Backhoe

Period	Cash Inflow	Present Value Factor	Present Value
1	$14,200	.8929	$12,679
2	14,200	.7972	11,320
3	14,200	.7118	10,108
4	14,200	.6355	9,024
5	14,200	.5674	8,057
	71,000		Total 51,188

Figure 6.6

Figure 6.7 provides the discount factors for the present value of an annuity. Since the cash inflows generated by the backhoe are consistent, this table can be used to determine the appropriate discount factor, in this case, 3.6048. This factor is the sum of the PV factors used above. The formula for the present value of an annuity is:

Present Value of an annuity =
Annuity × Present Value Interest Factor of an Annuity

Abbreviations for the formula variables are as follows.

Present Value of an Annuity = PVA
Annuity = A
Present Value Interest Factor of an Annuity = PVIFA

For Investment B, the solution is:

PVA $14,200 (12%, 5 periods)

= $14,200 × PVIFA (12%, 5 periods)
= $14,200 × 3.6048
= $51,188

These exercises are simple examples of discounted cash flow analysis used to evaluate the acquisition of capital assets. In the next section, we will discuss a variety of other factors that affect the discounted cash flow analysis and the capital budgeting process.

Present Value of an Annuity of $1 per Period

Number of Payments	2%	3%	4%	5%	6%	7%	8%	10%	12%
1	.98039	.97087	.96154	.95238	.94340	.93458	.92593	.90909	.89286
2	1.9416	1.9135	1.8861	1.8594	1.8334	1.8080	1.7833	1.7355	1.6901
3	2.8839	2.8286	2.7751	2.7232	2.6730	2.6243	2.5771	2.4869	2.4018
4	3.8077	3.7171	3.6299	3.5460	3.4651	3.3872	3.3121	3.1699	3.0373
5	4.7135	4.5797	4.4518	4.3295	4.2124	4.1002	3.9927	3.7908	3.6048
6	5.6014	5.4172	5.2421	5.0757	4.9173	4.7665	4.6229	4.3553	4.1114
7	6.4720	6.2303	6.0021	5.7864	5.5824	5.3893	5.2064	4.8684	4.5638
8	7.3255	7.0197	6.7327	6.4632	6.2098	5.9713	5.7466	5.3349	4.9676
9	8.1622	7.7861	7.4353	7.1078	6.8017	6.5152	6.2469	5.7590	5.3282
10	8.9826	8.5302	8.1109	7.7217	7.3601	7.0236	6.7101	6.1446	5.6502
11	9.7868	9.2526	8.7605	8.3064	7.8869	7.4987	7.1390	6.4951	5.9377
12	10.575	9.9540	9.3851	8.8633	8.3838	7.9427	7.5361	6.8137	6.1944
13	11.348	10.635	9.9856	9.3936	8.8527	8.3577	7.9038	7.1034	6.4235
14	12.106	11.296	10.563	9.8986	9.2950	8.7455	8.2442	7.3667	6.6282
15	12.849	11.938	11.118	10.380	9.7122	9.1079	8.5595	7.6061	6.8109
16	13.578	12.561	11.652	10.838	10.106	9.4466	8.8514	7.8237	6.9740
17	14.292	13.166	12.166	11.274	10.477	9.7632	9.1216	8.0216	7.1196
18	14.992	13.754	12.659	11.690	10.828	10.059	9.3719	8.2014	7.2497
19	15.678	14.324	13.134	12.085	11.158	10.336	9.6036	8.3649	7.3658
20	16.351	14.877	13.590	12.462	11.470	10.594	9.8181	8.5136	7.4694
25	19.523	17.413	15.622	14.094	12.783	11.654	10.675	9.0770	7.8431
30	22.396	19.600	17.292	15.372	13.765	12.409	11.258	9.4269	8.0552
40	27.355	23.115	19.793	17.159	15.046	13.332	11.925	9.7791	8.2438
50	31.424	25.730	21.482	18.256	15.762	13.801	12.233	9.9148	8.3045
60	34.761	27.676	22.623	18.929	16.161	14.039	12.377	9.9672	8.3240

Figure 6.7

Other Capital Asset Investment Factors

The acquisition of capital assets generally aims for an asset that will provide cash flow benefits to the firm for a number of years. In addition to the purchase price, the asset may require periodic cash investments, as in the case of equipment overhaul and maintenance. Whether the asset is financed with debt or by equity, there is still a cost for the financing. For *debt*, the cost is the explicit interest charged for the use of the funds. For *equity*, the cost is the opportunity cost of not having the funds available for other investments. Any investment of funds must generate a return sufficient to cover the cost of the funds.

The discounted cash flow technique most often used to evaluate capital asset acquisitions is *Net Present Value* analysis. In this procedure, the present value of cash inflows is compared to the present value of cash outflows over the life of the project. If the present value of the inflows exceeds the present value of the outflows, the acquisition would be accepted. The net present value (NPV) of the track loader in the previous example is $3,067; for the backhoe, it was $1,188. Since the projects were mutually exclusive with equivalent lives, the track loader should be chosen for acquisition.

Our discussion of capital budgeting has focused on *cash flows* rather than *accounting net income*. The reason for this focus is that it is important to understand not only the *dollar amount*, but also the *timing* of the cash flows. Typical cash inflows and outflows related to investments are as follows.

Cash Inflows:
- Additional revenue
- Labor savings
- Material savings
- Salvage value
- Release of working capital
- Tax savings from depreciation

Cash Outflows:
- Acquisition costs
- Repairs/Improvements
- Working capital commitment
- Taxes on profits

The contractor must make a good faith effort to forecast each of the above cash flow items in order to develop a good capital budgeting model.

Example 3

Eastway Contractors is considering the investment of $180,000 in a new piece of equipment. Purchase of the equipment will allow Eastway to avoid some equipment rental costs and will provide the opportunity to generate some additional revenue for the firm. The equipment is expected to last for eight years and have a salvage value of $25,000 at the end of the eight-year period. Eastway's controller is interested in whether or not the equipment will generate sufficient cost savings and additional revenue to cover the cost of purchase and operation. Before beginning the analysis, the controller forecasted the following.

Expected savings in annual equipment rental	$20,000
Annual revenue generated by the equipment	$40,000
Cash operating cost of the new equipment	($20,000)
Net increase in annual cash income	**$40,000**
Equipment overhaul in the fourth year	$10,000
Depreciation method - Straight Line	8 years
Eastway Contractors projected tax rate	30%
Discount Factor used to evaluate projects	**12%**

The following headings suggest a format useful for structuring the capital budgeting analysis:

Event   Years   Amount   PV Factor   Present Value

These headings are briefly defined below.

Event: The transaction that causes a cash outflow or inflow.

Years: The number of annual periods affected by the transaction. Some events (e.g., the initial purchase) occur only once, while others (e.g., annual cost savings) occur over a period of years.

Amount: The after-tax cash inflow or outflow associated with the transaction.

PV Factor: The factor, as specified in Figures 6.5 and 6.7, which converts the future cash inflow or outflow to present value dollars.

Present Value: The product obtained by multiplying the Amount by the PV Factor.

Figure 6.8 depicts the completed capital budgeting analysis. The initial event, equipment purchase, represents the major cash outflow of the project. It occurs in Period 0 (the current period) and represents an outflow of $180,000. For simplicity, it is assumed in this example that the firm pays the entire amount, up front, using working capital. In reality, it is much more likely that the firm would make an initial downpayment only, with additional monthly payments scheduled over a period of several years.

Annual Cash Income

The next event is the annual increase in cash net income due to the acquisition of equipment. It is important to note that the contractor does not have access to the entire increase in cash net income. This is because the increase in net income causes an increase in the income taxes that must be paid. The increase in cash net income must, therefore, be offset by the increase in taxes. On the schedule, this tax adjustment has been calculated by multiplying the annual increase in cash net income by one, minus the firm's tax rate, which is 30% (1–30% = 70%). The result is an after-tax increase in net income of $28,000. In essence, Eastway keeps 70% of the profits and the taxing authorities receive the other 30%. This annual income represents a cash inflow for the firm for the next eight years and, therefore, is multiplied by the PVIFA (eight periods at 12% shown in Figure 6.7) of 4.9676. The result is a present value of cash inflows of $139,093.

Depreciation

Depreciation is the allocation of an asset's cost over its useful life. Depreciation is not a cash expense; therefore, it does not require a direct cash disbursement. The entire cost of the equipment has already been considered as a cash outflow. The indirect effect of depreciation is as a deduction on the firm's tax return, where it provides a tax savings and therefore a cash savings.

The annual depreciation amount for our example is $22,500 (180,000 divided by 8). This deduction on the tax return will save the firm $6,750 (22,500 × .30) in income taxes. Since this amount is a very real cash savings, it represents a cash inflow to the firm over the next eight years. Using the same Present Value Factor as that applied to the annual cash net income, a Present Value of $33,531 (6,750 × 4.9676) is calculated for the cumulative depreciation tax savings over 8 years at 12%.

Overhaul: The next event noted is the overhaul of equipment, scheduled to take place at the end of the fourth year. The $10,000 overhaul represents a cash outflow for the firm. It is, however, a tax deduction (assuming a full write-off of the amount) and, therefore, will generate tax savings. The net amount of cash outflow resulting from the overhaul is calculated by multiplying the cost of the overhaul by one, minus the firm's tax rate, which is 30%. The result is a net cash cost of $7,000 (10,000 × .70). This amount multiplied by .6355 (the PVIF $1 for Period 4 at 12% shown in Figure 6.5) results in a present value for the cost of the overhaul of $4,449.

Capital Budgeting Analysis

Event	Years	Amount	PV Factor	Present Value
Equipment Purchase	0	($180,000)	1.000	($180,000)
Annual Cash Net Income $40,000 (1–.30)	1–8	28,000	4.9676	139,093
Depreciation Tax Savings $22,500 (.30)	1–8	6,750	4.9676	33,531
Year 4 Overhaul $10,000 (1–.30)	4	(7,000)	0.6355	(4,449)
Year 8 Salvage Value $25,000 (1–.30)	8	17,500	0.4039	7,068
			Net Present Value	($4,757)

Figure 6.8

Salvage Value: The final event affecting the investment in equipment is the salvage value of the equipment at the end of the eighth year. For purposes of analysis, it is assumed that the equipment will be sold at the end of the eighth year for the $25,000 salvage value. Since the equipment is fully depreciated at that point, the entire $25,000 is a taxable gain. Since the firm's tax rate is 30%, the remaining 70% is available to the firm as a cash inflow in the eighth year. The $17,500 ($25,000 × .70) cash inflow is discounted back to the present using a factor of .4039. This factor is the PVIF for an amount received in Period 8 at 12% (Figure 6.5) and results in a present value of $7,068.

Deciding on the Investment: The net present value of the cash inflows and outflows is a negative $4,757. This figure indicates that the investment does not generate the required 12% return and, therefore, should not be accepted unless other considerations warrant it. The analysis does not disclose the *actual rate of return* on the investment. The actual rate can be approximated by substituting a lesser rate of return in the foregoing analysis and, by trial and error, discovering the rate of return closest to making the net present value of the project equal 0. There are mathematical procedures that can be used to determine actual rate of return, but the net present value approach yields adequate results for the purpose of comparing alternative investments.

Other Approaches to Capital Budgeting Decisions: Some managers prefer to use one of the other general approaches to capital budgeting. The reason is usually ease of use. One of these other approaches is the *Payback Method*. This method focuses on the length of time it takes an investment to recover its initial cost. For example:

Eastway Contractors needs a new machine and is considering two options, Machine X and Machine Y. Machine X costs $20,000 and Machine Y costs $18,000. Machine X will generate annual cash inflows of $5,000 and Machine Y will generate annual cash inflows of $4,600. The calculation of the payback period for each machine is as follows:

$$\text{Machine X} \frac{\$20,000 \quad \text{(initial cost)}}{\$5,000 \quad \text{(annual cash inflow)}} = 4.0 \text{ years to recover cost}$$

$$\text{Machine Y} \frac{\$18,000 \quad \text{(initial cost)}}{\$4,600 \quad \text{(annual cash inflow)}} = 3.9 \text{ years to recover cost}$$

Since Machine Y has a faster payback, it would be selected based on the payback method criteria.

The payback method has two primary flaws. First, it does not consider cash flows after the payback period and, second, it does not address the time value of money. For short term or very risky projects, the payback method has particular merit and should be used in conjunction with the discounted cash flow method previously discussed.

Capital budgeting techniques represent a *quantitative* process, helpful to the manager in making investment decisions. There are many options for the contractor's investment dollar. It is the manager's goal to direct available dollars to the investment that is expected to produce the *most significant benefit* for the firm. While none of us can predict the future with absolute certainty, careful consideration of the different benefits and costs of each alternative will result in a better ultimate decision.

General Management Procedures: One of the major factors in the decision to acquire equipment is the projection of need for the piece of equipment. The contractor must be able to separate a "nice to own" piece of equipment from a "need to own" piece of equipment. After considering all factors, the contractor may decide that hiring subcontractors with their own equipment or renting the equipment is more economically feasible than purchasing the equipment. The final decision should be the result of a *quantitative analysis* and *general management review*.

The acquisition of equipment brings with it other associated costs that must be considered by the contractor in the decision-making process. For example, *storage and insurance costs* will be incurred on the equipment even if it is not in use. Certain pieces of equipment may also require skilled operators. In any event, the cost of "down time" must be considered as an associated cost of equipment purchase. Other major considerations include:

1. The age, maintenance costs, and efficiency of equipment currently owned.
2. Warranties provided with new equipment.
3. The outlook for new projects and the economy in general.
4. *All* costs of ownership. Have they been considered?
5. Alternative arrangements. What terms are offered by other leasing agencies?
6. Inflation. Is it likely to affect the future purchase price of the equipment?
7. Future tax changes. Are there any revisions on the horizon that will affect the net return on investment?

The purchase of a capital asset is only the beginning of the asset management process. Since capital asset expenditures represent a significant obligation of funds, proper care and monitoring of the use of the asset must be ensured.

It is important to maintain detailed records regarding the acquisition of capital assets. The records should include the following.

- Asset Description
- Cost
- Purchase Date
- Model/Serial Number
- Warranty Information
- Seller

This information is needed to maintain current records of the assets owned by the contractor, for such purposes as:

- Proper depreciation expense calculations
- Internal control over equipment security
- Proper calculation of gain or loss upon disposal
- Security liens by financial institutions

Summary

Asset management focuses on the effective use of the contractor's assets in meeting the objectives of the firm. Each asset plays an important part in the overall success of the firm, and thus requires (to varying degrees) management's attention. Cash, receivables, inventory, and plant and equipment are the major assets employed by the contractor and, therefore, merit their own specific policies and procedures.

Current Asset Management

Current asset management concerns those assets that will be converted into cash or used up within one year. The primary emphasis in current asset management is the efficient and timely turnover of the asset. Problems such as old accounts receivable and slow moving inventory should be minimized.

The contractor's accounts receivable management should be supported by a sound credit policy. The credit policy should cover the following key areas.

1. Credit Standards
2. Credit Terms
3. Credit Account Tracking
4. Collection
5. Timely Billing Procedures

Capital Asset Management

Capital asset management involves the management of assets that will provide the firm with benefits for periods beyond one year. Typical capital asset management issues include:

1. Equipment Expansion/Replacement Decisions
2. Plant Expansion/Improvement Decisions
3. Purchase vs. Lease Decisions

Since capital asset decisions involve the acquisition of an asset that will affect future cash flows, either in a single period or a series of periods, the contractor must consider the *time value of money*. Because of the erosion of the purchasing power of the dollar, a dollar received today is worth more than a dollar received in the future. Thus the contractor must consider not only the *amount of* the cash flows, but their *timing*.

Chapter Seven

DEBT MANAGEMENT

Chapter Seven

DEBT MANAGEMENT

The purpose of this chapter is to examine the different types of financing available to the contractor. The balance sheet for our hypothetical contracting firm, Eastway Contractors, shows that the firm's total assets equal its total liabilities and net worth. (The balance sheet appears in Chapter 3, Figure 3.2.) In essence, the liability and net worth sections of the balance sheet show how the contractor has managed to finance the assets used for the firm's operations. The focus of this chapter is the *liability* section of the balance sheet.

The chapter is divided into two main sections: *short term credit* and *long term credit*. The maturity structure of a firm's debt should, in part, match the structure of the firm's asset base. Generally, short term debt should finance current assets, and long term debt should finance capital, or long term assets.

Short Term Credit

Short term credit is defined as any liability that must be satisfied within one year. The major sources of short term credit for the contractor are accruals (such as wages and taxes owed but not yet paid), trade credit, and short term loans from lending institutions.

Accruals

Accruals are the spontaneous increases in liabilities due to normal ongoing business activity. Each day that the contractor's employees work for the firm, they are earning wages, yet the firm does not pay the employees on a daily basis. Payroll is generally disbursed every week or every two weeks. Up until that point, the employees' wages are *accruing*. The liability is satisfied when payroll is disbursed. Even on payday, the full amount of the earned wages is not actually paid, since the payroll tax withholdings are retained by the employer until payment is due to the taxing authorities.

Income taxes are another example of a liability accrual. As the firm generates taxable net income throughout the year, the liability for income taxes is accruing. The amount is not paid until the firm's quarterly filing of estimated taxes. Accruals provide a source of "free" credit, since the firm enjoys the use of the resource, but is able to postpone payment for the resource.

There is a very tight limit on the use of accruals as resources for financing since the cost for late payment (especially in the case of payroll taxes) can be very high. Accrual financing is, therefore, a form of free credit that should be used extensively, but repaid in a timely manner.

Trade Credit

Trade credit is probably the most common form of short term financing used by the contractor. It is the ongoing accounts payable that have been established with suppliers, subcontractors, and other vendors of the firm. For smaller firms, trade credit may represent a larger portion of the overall short term financing, since these firms may not yet be able to qualify for bank financing. Like accruals, trade credit is thought of as an ongoing, "spontaneous" source of financing. As the contractor's business expands and the purchase of materials and services from vendors increases, the amount of the accounts payable will automatically increase.

The terms for trade credit vary from industry to industry, but net 30 days is very common. Thus, if the contractor is making purchases of $1,000 each day and the payment period is extended to the entire 30-day period, the average accounts payable balance will be $30,000. If the payment period is lengthened to 40 days, then the average balance will increase to $40,000. Likewise, if the contractor increases purchases from $1,000 to $1,500 per day, the average balance for accounts payable will increase to $45,000. No explicit action was taken to increase the level of trade credit; its growth is simply the result of increased business activity. The contractor's debt financing is thereby increased, and for a short time, it is a source of free financing.

Since trade credit is a primary source of financing for many firms, it is important to maintain strong credit relationships with the firm's vendors. Often, when a contractor is unable to arrange bank credit because of lack of business history or volatile business conditions, trade creditors will step in and extend credit for purchases. Generally, creditors in the industry understand the contractor's position and business conditions better than a traditional lending institution. Vendors also serve as references for future credit relationships and customer referrals. This important relationship should not be endangered by delinquent and inconsistent payment practices. By the same token, while it is in the vendors' best interest to extend credit in order to promote the sale of their products, they are careful not to overextend credit to suspect customers.

One of the advantages of maintaining strong relationships with vendors is that in times of real need, the vendor may be more willing to consent to extended payment terms. When restricted cash flow leads to late payment on vendor invoices, it is best to advise the vendor of the pending late payment, and provide a time frame for the full payment of the obligation. Once this time frame is established, it is very important to stick to it, meeting all deadlines. Credit is often extended based on the vendor's confidence in the contractor, even when the numbers may not quite justify the extension. Maintaining the vendors' confidence is extremely important.

Credit Terms: Credit terms (discussed in detail in Chapter 6) are the payment policies created by the vendor as part of a credit transaction. The terms for the payment period, discounts, and finance charges may vary widely from vendor to vendor. Generally, vendors in the same industry have similar credit terms in order to remain competitive.

If the vendor is in a "tight" cash position, the terms might reflect discounts for early payment and substantial penalties for late payments. In some industries, payment terms are unusually long, in order to allow the purchaser some lead time before satisfying the obligation. For example, the farm equipment industry is known for providing extended deferral of initial payments. Having this extra time allows the farmer to generate income from the resource (equipment) prior to payment. Some manufacturers go so far as to ship a product to a customer and provide liberal payment terms just to entice the customer to accept the product and relieve the manufacturer of the inventory carrying cost. Other industries, particularly those whose products are in high demand, provide very short, if any, payment terms.

If credit terms offered by vendors seem "too good to be true", be wary of inflated prices or inferior goods. Generally, one firm will not provide terms substantially different from other firms in the industry. It does happen, but it is unusual.

Early Payment Discount

In addition to the length of time allowed for payment, the vendor may specify a cash discount that can be deducted from the purchase price if the customer makes early payment. Cash discounts are offered to encourage the purchaser to make payment within a specific number of days. The following format is typically used to specify terms: 2/10, net 30.

This particular set of terms specifies that if payment is made by the tenth day after the invoice date, the purchaser may deduct a 2% discount from the stated invoice amount. If payment is not made by the tenth day, the full amount is due by the 30th day following the date of the invoice. Two percent may not sound like a substantial amount, but the annualized cost of not taking advantage of the early payment terms is considerable.

If the contractor fails to take advantage of the discount and pays the full invoice amount within the 30-day period, the annualized cost of the financing would be 37.23%. By failing to take advantage of the discount, the contractor indicates that he is willing to pay two percent of the invoice amount for the right to use the remaining 98% of the purchase price for a 20-day period. The following formula may be used to calculate the effective annual cost of failing to take cash discounts.

$$\frac{\text{Discount Percent}}{100\% - \text{Discount Percent}} \times \frac{365}{\text{Payment Period} - \text{Discount Period}}$$

For the 2/10, net 30 example, the calculation would be:

$$\frac{2\%}{100\% - 2\%} \times \frac{365}{30 - 10} = .0204 \times 18.25 = 37.23\%$$

The following are other examples of effective annual rates resulting from failure to take advantage of offered cash discounts. These figures are based on three different net terms.

Terms	Effective Annual Rate
1/10, net 30	18.43%
1/10, net 15	73.73%
2/10, net 20	74.46%

These rates are much higher than the cost of financing through commercial banks or other lending institutions. Unless this form of trade credit is being used as the financing of last resort, the manager should always try to arrange other short term financing to cover cash needs and take advantage of cash discounts.

The effective cost of trade credit can be altered by deviating from the stated credit terms. For example, paying a 2/10, net 30 invoice in 40 days instead of 30 days will reduce the effective cost of financing to 24.8%. This practice may work on occasion, but constant delinquent payments will damage the contractor's credit rating and endanger subsequent trade credit relationships.

More often than not, firms fail to take advantage of cash discounts, not because of cash shortages, but due to inefficient accounts payable tracking. Many firms benefit from a chronological filing system for all accounts payable, based on the due date of each invoice. Using this system, a cash requirements report should be generated weekly, specifying the cash needs for the following week. Management then has the opportunity to plan for upcoming cash requirements, and being prepared, can take advantage of pending cash discounts. This "planning window" also allows the contractor to effectively plan the debt financing needed to cover cash shortages.

Line of Credit

Arranging a *line of credit* at the contractor's bank is a good alternative to costly trade credit. It makes good business sense to pay the 10–12% line of credit rate, rather than the 36% trade credit rate. This argument makes sense to the banker as well.

A line of credit is a financing arrangement wherein the contractor has pre-established authority to borrow up to a specified dollar amount. The ceiling on the borrowing limit is based on a combination of factors including the contractor's collateral and ability to repay the loan in a timely manner, the firm's cash flow position, and the stability of the firm as well as that of the industry. Remember, the focus here is on the contractor's ability to repay short term debt. This criteria may be contradictory to the company's long term growth pattern and profitability.

Typically, the contractor requests a line of credit in order to finance the working capital requirements of major expansion or new project start-up. It is important to understand up-front the requirements for loan repayment. If the line of credit is designed for short term financing, then the contractor should not obligate the funds to financing long term assets.

Many times, a line of credit has a stipulation requiring the borrower to maintain a zero balance for a consecutive number of days during the year, typically 30. With this requirement, the lending institution ensures that the borrower is not obligating the funds to long term financing. There are many cases where a borrower agrees to the terms in a line of credit and then is unable to repay the amount in time to adhere to the 30-day zero balance requirement. Such stipulations are not included in all line of credit arrangements, but they are very common. If the line of credit does not specify a zero balance during some portion of the year, then a minimum monthly payment may be required. Generally, the payment is a percentage of the outstanding balance and changes as the balance changes. In most cases, there should be no prepayment penalty for accelerated payments against the principal.

Interest on the line of credit is based only on the funds actually borrowed. However, the lending institution may assess a *commitment fee* on the unused portion of the line of credit. The position of the lending institution is that the amount of the line of credit is available for the borrower's use and thus, that portion of the institution's lending capacity is unavailable for use by other customers. The commitment fee is generally in the range from one half to one percent and is only assessed if the lending institution has made a firm commitment to obligate the funds. Payment of a loan commitment fee has the effect of raising the effective interest rate of the loan.

At times, the lending institution may require the borrower to maintain a *compensating balance*. A compensating balance is an amount, usually a percentage of the outstanding loan balances, that the borrower must maintain on deposit in the lending institution. The amount may be on deposit in an interest bearing or non-interest bearing account. The compensating balance (especially if it is in a non-interest bearing account) has the effect of raising the effective interest rate of the loan.

Interest rates charged by lending institutions vary. As a general rule, one percent above the bank's prime rate is considered average, one half percent above prime is good, and a rate equal to prime is excellent. Only the most stable and well managed construction firms can expect to borrow on a line of credit at the bank's prime rate.

The contractor's goal should be to create as flexible a relationship as possible with the bank. The fewer strings attached to the line of credit, the better. Ideally, the contractor wants immediate access to the funds for any reason, with minimal explanation.

The lending institutions use a variety of approaches to collateralize lines of credit. For example, the balance of the contractor's billed receivables and unbilled work-in-process might be used as loan security. In this case, the contractor would be required to prepare a schedule each month, detailing the outstanding accounts receivable and the investment in work-in-process. Based on the balances, the contractor may be able to draw down additional funds from the line of credit, *or* may be required to repay previously borrowed funds.

To illustrate this approach to a line of credit, suppose Eastway Contractors establishes a relationship with First Bank whereby Eastway is granted a $200,000 line of credit. The line of credit is secured by accounts receivable and work-in-process. At the end of each month, Eastway is required to furnish the bank with a current schedule of accounts receivable and work-in-process. According to the loan agreement, the amount of the line of credit draw down may be limited to 60% of the balance of the collateral. Therefore, if the total of the accounts receivable and the work-in-process equals or exceeds $333,333, then Eastway would have access to the entire $200,000. If the balance falls below $333,333, then Eastway would only be entitled to 60% of the collateral amount.

This type of financing, called *asset based lending*, is generally an advantageous arrangement for the contractor. More funds become available as business expands (until reaching the "ceiling" on the line of credit). Available funds decrease as business slows down. Many smaller lending institutions will not enter into this type of arrangement due to the bookkeeping requirements. Those that do must have a knowledgeable staff capable of both monitoring the reports submitted by the contractor and making on-site inspections of the work-in-process. Many lending institutions find that these requirements exceed their basic operations and would rather have fixed assets serve as collateral. Buildings, land, and equipment are the types of assets preferred by most lenders. In most financing arrangements, especially for smaller firms, the owner's signature and personal guarantee are almost an absolute requirement.

Accounts Receivable and Inventory Financing

Accounts receivable and inventory can also be used to secure financing, as they are used for collateral in lines of credit. Accounts receivable financing involves either *pledging* or *factoring* the accounts receivable.

Pledging: With pledging, the contractor uses accounts receivable as security for a loan. The lending institution has recourse against the contractor in case the contractor's customer does not pay the account receivable when due. Using this method, contract receipts are assigned to lending institutions as security for loans. The general contractor signs a release authorizing the assignment of contract proceeds. Subcontractors must be aware of any clauses in the contract that allow such assignment of the proceeds. This can be a sensitive issue, however, particularly when the relationship between the subcontractor and the general contractor is new.

Factoring: Factoring involves the outright purchase of the contractor's accounts receivable by the lending institution. This purchase may or may not provide recourse to the lending institution. Generally, the contractor's client is notified of this arrangement and is instructed to make payment directly to the lending institution. If the accounts are purchased without recourse, the lending institution assumes the risk of default by the customer.

Although accounts receivable financing can be expensive, it does provide some benefits to the firm. It accelerates the collection of cash, thus reducing the firm's cash conversion cycle. Furthermore, it may be more cost effective for the contractor to sell the receivables to a lending institution with a well established credit department, rather than to incur the cost of setting up an in-house credit department. With advancements in computer technology, the ability of lending institutions to check customer credit ratings is enhanced. As a result, these institutions are more willing to enter into accounts receivable financing arrangements.

Short Term Loans

Short term loans are similar to lines of credit, except that the amount of the debt is fixed at the time that the borrowing and repayment schedule is fixed. For short term loans, the principal is often repaid in a lump sum at the end of the note term. Interest is usually paid on a monthly basis. Collateral requirements are also similar to those for the line of credit. Once again, it is important to note that the financing is short term in nature and should not be used to finance fixed assets. A short term loan is typically used to answer a very specific financing need when the contractor does not want to obligate the more flexible line of credit.

Bank Loans

Bank loans are probably the most important source of short term financing for the contractor. The primary source of short term bank financing is the commercial bank. However, savings and loan institutions also make such loans. Bank loans differ from accruals and trade credit in that they cannot be considered spontaneous or "automatic" financing, since explicit action must be taken to arrange the bank loan. If this type of financing cannot be arranged, the contractor may have to curtail operations or growth.

Short term bank loans can be arranged in different forms. The two primary types of short term loans are a *term* and a *line of credit* loan. Each of these has characteristics similar to all debt instruments.

Interest Rates: Calculating the Cost of Financing

The cost of a short or long term bank loan depends upon a number of factors, including the riskiness of the borrower, the size of the loan, the level of interest rates in the marketplace, and the term of the loan. Interest rates are also a function of three major variables:

- the inflation rate
- a real rate of return for the lender
- a risk factor

The contractor cannot do anything about the inflation rate or the rate of return required by the lending institution. Some control can, however, be exercised over the perceived riskiness of the firm. Factors that increase a perception of risk include a high debt ratio, a significant amount of old accounts receivable, delinquent payments on payroll taxes and estimated tax payments, general decline in business activity, and financial problems among major customers. While the contractor cannot control all of these factors, it is important to focus on maintaining good management over the areas that can be controlled.

Interest rates over the past ten years have been very volatile. Fortunately, the more recent years have shown fairly steady rates and often favorable declines. The impact of interest rates on construction activity, and therefore contractors, is well documented.

To help protect themselves against volatile interest rates, most institutions now use *variable rate loans* instead of *fixed rate loans*. A variable rate loan is one in which the interest rate changes periodically (monthly, quarterly, etc.), based on the change in an established economic index. For example, the lending institution's interest rate might be tied to the *prime* interest rate (that rate given to the strongest customers) and quoted at 1% over prime, to be adjusted monthly. At the end of each month, the lending institution examines the level of the prime interest rate and adjusts the loan interest rate accordingly. If the prime rate moves up, the loan rate also moves up; if the prime rate drops, the loan rate drops.

The variable rate loan clearly has advantages for the bank, but it could put the contractor into a bind. The prime interest rate has a direct impact on the business activity of contractors. If interest rates increase, there is usually a curtailment of building and capital asset expansion. Unfortunately, at the same time this curtailment is taking place, the interest rate on the contractor's debt is increasing. The increasing interest rate requires a larger payment and greater demands on cash flow at a time when business may be slowing down. This may restrict the contractor's cash flow.

While contractors are at risk to a high-interest rate situation, they also stand to receive a two-fold benefit should interest rates decline. With lower interest rates, not only does the contractor's debt service decrease, but the level of building and capital expansion usually increases. This translates into increased business for the contractor and potentially higher profits. Thus, declining interest rates mean a reduced cash demand on the contractor, and a growth in his business. Unfortunately, the ability to accurately forecast interest rates over an extended period is rare and usually a matter of luck. A study of trends can help the contractor position the firm to reduce exposure to unfavorable interest rate changes. If indications are that interest rates are on the rise, the contractor may want to begin "locking up" longer term fixed rate debt instruments. One of the advantages of fixed rate debt is that it reduces the uncertainty of cash flow planning and the ability to cover debt.

Calculating the Cost of Financing

Contractors should have an understanding of the methods used to calculate the effective annual cost of various financing alternatives. The following section is a discussion on calculating the effective cost of short term loan financing.

The basic calculation for interest is:

$$\text{Interest} = \text{Principal} \times \text{Rate} \times \text{Time}$$

Principal refers to the amount borrowed; *rate* is the effective annual interest rate; and *time* is the length of the note term in years.

Simple interest means that the interest charge is calculated on the principal amount only. The effective cost of simple interest is calculated as follows:

$$\text{Effective rate of interest} = \frac{\text{Interest}}{\text{Principal}}$$

For example, on a $50,000 loan for one year at 10%, the interest is calculated as follows:

$$\$50,000 \times .10 \times 1 = \$5,000$$

The effective rate of interest is calculated as:

$$\text{Effective rate of interest} = \frac{\$5,000}{\$50,000} = 10.0\%$$

Suppose the lending institution discounts the loan? When the bank deducts the interest in advance, it is called *discounting the loan*. The effect of discounting the loan is to increase the effective interest rate, since the borrower is receiving the principal amount less the interest charge. Calculating the effective interest rate on a discounted version of the same loan is as follows:

$$\text{Effective rate of interest} = \frac{\$5,000}{\$45,000} = 11.11\%$$

The amount of the principal of $50,000 less the $5,000 interest cost is $45,000. The effective interest rate increases by 1.11%.

In the next example, the lending institution is offering a simple interest loan at 10%, but requires that a 15% compensating balance of $7,500 be maintained in the contractor's non-interest bearing checking account. Therefore, the contractor only has 85% of the principal amount available since the remainder (15%) constitutes the required compensating balance. The effective interest rate on a loan with a compensating balance requirement is calculated as follows:

$$\frac{\text{Effective rate}}{\text{of interest}} = \frac{\text{Interest charge}}{\text{Principal amount less the compensating balance}}$$

$$= \frac{\$5,000}{\$42,500} = 11.76\%$$

The $42,500 is equal to the $50,000 principal amount less the incremental funds that must be on deposit in the contractor's non-interest bearing checking account. Fifteen percent of $50,000, or $7,500, is the incremental balance.

In another example, a lending institution imposes a 1% loan fee up-front on a simple interest loan. The loan fee increases the effective cost of the loan. The effective interest rate is calculated as follows:

$$\text{Effective rate of interest} = \frac{\text{interest charge plus loan fee}}{\text{loan principal}}$$

$$\frac{\$5,500}{\$50,000} = 11.0\%$$

The $5,500 amount is equal to the $5,000 interest charge plus the $500 loan commitment fee.

The effective cost of one type of loan must be calculated and compared to the effective cost of other types of loans. This is the only way to obtain a true comparison of the cost of financing.

Long Term Debt

It usually takes several years before an investment in long term or capital assets generates enough income to pay for itself or, in many cases, to even become profitable. During this waiting period, the investment is not likely to generate sufficient cash to allow for total repayment of the initial financing. For this reason, long term capital assets should, in almost all cases, be financed with long term, rather than short term debt. Even if the short term debt can be renewed at the same rate each year, the degree of risk increases materially when long term assets are financed with short term debt obligations. The risk is based on the unpredictability of the interest rates at the time that the loan is renewed. The interest rates may be high, or the contractor's financial position may be at a point where it does not meet the bank's criteria for renewing the loan.

The process of matching debt with the assets being financed is called *maturity matching*. Current assets should generally be financed with short term financing, and long term capital assets with long term financing. In this way, the contractor can match the conversion of the asset into cash with the repayment of debt. This arrangement assumes that the contractor has a permanent working capital investment, at least until the retained earnings of the business can support the permanent working capital needs.

Long term debt is advisable when the following conditions exist.
- Inflation is expected to increase, thus causing a corresponding increase in interest rates.
- The contractor has unused debt capacity as measured by the firm's debt ratio.
- A substantial benefit is expected from the use of leverage.
- Primary asset expansion will be in the form of capital assets, rather than current assets.

Term Loans

Long term debt is generally structured as a *term loan*, an agreement under which the borrower makes a series of interest and principal payments on specific dates to the lender. Generally, term loans are *amortized*, or paid off in a series of equal installments. Figure 7.1 is a partial amortization schedule for a loan of $100,000 for ten years at 10%. Note that each payment is applied against both the principal and the interest, with the portion applied to the principal increasing with each payment. The table in Figure 7.2 can be used to calculate the monthly payments required to amortize a loan at various interest rates and years.

Loan Amortization Schedule

Parameters:

Initial Principal	100,000.00
Interest Rate	10.00%
Loan Term (Years)	10
Payments Per Year	12
Payment Amount	1,321.51

Payment Number	Payment Amount	Interest Amount	Principal Reduction	Remaining Principal
0	—	—	—	$100,000.00
1	$1,321.51	$833.33	$488.17	99,511.83
2	1,321.51	829.27	492.24	99,019.58
3	1,321.51	825.16	496.34	98,523.24
4	1,321.51	821.03	500.48	98,022.76
5	1,321.51	816.86	504.65	97,518.11
6	1,321.51	812.65	508.86	97,009.25
7	1,321.51	808.41	513.10	96,496.15
8	1,321.51	804.13	517.37	95,978.78
9	1,321.51	799.82	521.68	95,457.10
10	1,321.51	795.48	526.03	94,931.07
11	1,321.51	791.09	530.42	94,400.65
12	1,321.51	786.67	534.84	93,865.82
13	1,321.51	782.22	539.29	93,326.52
14	1,321.51	777.72	543.79	92,782.74
15	1,321.51	773.19	548.32	92,234.42
16	1,321.51	768.62	552.89	91,681.53
17	1,321.51	764.01	557.49	91,124.04
18	1,321.51	759.37	562.14	90,561.90
19	1,321.51	754.68	566.82	89,995.07
20	1,321.51	749.96	571.55	89,423.52
21	1,321.51	745.20	576.31	88,847.21
22	1,321.51	740.39	581.11	88,266.10
23	1,321.51	735.55	585.96	87,680.14
24	1,321.51	730.67	590.84	87,089.30

Figure 7.1

The total interest cost and equity buildup in a 30-year amortization is so significant that reducing the note to a 15-year amortization at a 10% interest rate changes the monthly payment by only approximately $2.00 per $1,000 borrowed. A borrower could pay off a 30-year $50,000 loan in one-half the time (15 years) by increasing the monthly payment by only $100. The contractor should consider the overall cost of financing when selecting the most advantageous debt structure. If $100 per month is a critical amount and is the difference between positive and negative cash flow, then the contractor may need to extend the payment term.

Monthly Payment Needed to Amortize a Loan of $1,000

To calculate your specific payment, multiply the amount shown on the schedule times the number of thousands being borrowed. For example, if you borrow $20,000 at 10% for 10 years, you would multiply 20 times $13.22 to calculate the monthly payment ($264.40). $60,000 at 9% for 15 years would be = 60 times $10.14 ($608.40).

Loan Term (Years)	Interest Rates							
	8.50%	9.00%	9.50%	10.00%	10.50%	11.00%	12.00%	13.00%
1	87.22	87.45	87.68	87.92	88.15	88.38	88.85	89.32
2	45.46	45.68	45.91	46.14	46.38	46.61	47.07	47.54
3	31.57	31.80	32.03	32.27	32.50	32.74	33.21	33.69
4	24.65	24.89	25.12	25.36	25.60	25.85	26.33	26.83
5	20.52	20.76	21.00	21.25	21.49	21.74	22.24	22.75
6	17.78	18.03	18.27	18.53	18.78	19.03	19.55	20.07
7	15.84	16.09	16.34	16.60	16.86	17.12	17.65	18.19
8	14.39	14.65	14.91	15.17	15.44	15.71	16.25	16.81
10	12.40	12.67	12.94	13.22	13.49	13.78	14.35	14.93
12	11.10	11.38	11.66	11.95	12.24	12.54	13.13	13.75
15	9.85	10.14	10.44	10.75	11.05	11.37	12.00	12.65
18	9.05	9.36	9.68	10.00	10.32	10.65	11.32	12.00
20	8.68	9.00	9.32	9.65	9.98	10.32	11.01	11.72
25	8.05	8.39	8.74	9.09	9.44	9.80	10.53	11.28
30	7.69	8.05	8.41	8.78	9.15	9.52	10.29	11.06

Figure 7.2

Dealing with Lending Institutions

The Contractor's Credit Image

The contractor's ability to obtain credit is largely dependent on the extent to which he understands and is able to respond to the concern of lenders and sureties. The image the contractor projects in terms of management, profitability, and internal control is crucial to the firm's relationships with these creditors. Creditors are clearly more willing to support a contractor who has a track record of managing a financially solvent and profitable company.

Certain negative conditions represent particular concerns to creditors who wish to establish a flexible relationship with a construction firm. The most significant factors are as follows:

1. **Unprofitable Jobs:** A history of losses on past projects suggests potential problems with estimating, supervision, subcontractors, or disputes with owners or architects. The creditor will often investigate the firm's ability to estimate and price jobs accurately by reviewing job cost and job status reports. The sophistication of the job cost system and its ability to alert management promptly to shortfalls is also of great concern, for if management lacks current knowledge of mounting losses, it will be difficult for the creditor to acquire this knowledge.

2. **Poor Receivables:** Consistently slow or uncollectable receivables may indicate poor contractual protection, insufficient credit policies, disputes, or simply poor management.

3. **Improper Volume:** Inadequate volume will result in a level of gross profit incapable of covering the firm's operating expenses, much less yielding a profit. Excessive volume may strain the firm's management control systems and create cash flow shortages. Creditors will likely be interested in knowing the portion of operating expenses that will be covered in future periods by current uncompleted contracts (backlog) versus new projects that are not yet under contract.

4. **High Operating Expenses:** Most creditors compare the firm's operating expenses to industry averages. Unusually high expenses may indicate the firm is top-heavy with management, or simply inefficient. Often, the problem is excessive salaries paid to the owner or top management, reducing both profit and working capital.

5. **Excessive Capital Assets:** Many contractors make the fatal mistake of investing too heavily in such capital assets as buildings and equipment. Independent of the inherent economics, this approach may reduce working capital initially, require excessive interest payments, and strain the firm during slow periods. Creditors will certainly be concerned with the debt obligations to which the firm has already committed, and the terms of those obligations.

6. **Real Estate:** Creditors are generally not pleased with contracting entities that invest in real estate developments. The financial risk of such investments and the potential strain on working capital are great concerns. Such investments should be made on a personal basis.

7. **Unfamiliar Construction:** While diversification can be crucial to the success of the construction firm, it must be approached with caution. Many firms have gone bankrupt while trying to enter a new line of construction or new geographical area. Unfamiliar techniques, contractual requirements, or local standards may cause immediate problems.

8. **Officer's Receivables:** The owners of many construction firms borrow funds from their companies on a regular basis. As long as the amount is limited and it is borrowed only on a short term basis, this practice may be acceptable. However, when such borrowings dig too deeply into needed working capital and such loans extend beyond a year (perhaps never to be repaid), this practice becomes a serious concern to creditors.

9. **Financial Reporting Practices:** Creditors are not impressed with financial statements generated using the cash and completed contracts methods. These methods can be very misleading, as explained in Chapter 3. Creditors will also be concerned with the accounting for taxes, leased equipment, overbillings, inventory valuation, allocation procedures for indirect costs, and accrued obligations for fringe benefits.

10. **Litigation:** Of course, any legal exposure arising out of litigation or arbitration is of immediate concern to any creditor. If such exposure does exist, it may be desirable to provide the creditor with a written explanation from the firm's attorney regarding the situation and likely outcome.

Construction-wise creditors will investigate the existence of all of these conditions and develop their relationship with the contractor accordingly. If the contractor is exposed on any of these points, it is best that he prepare a believable written explanation of the circumstances and include it in the loan application. Likewise, any shortfall in financial structure or performance, as compared to industry averages, should be addressed up-front. A further discussion on the loan application process is included later in the chapter.

Selecting a Lender

When selecting a lending institution, the contractor should consider the following factors.

- Banks have varying philosophies toward particular industries. Some banks make a concerted effort to develop an *expertise* in a particular industry. This enables the bank to respond more effectively to the needs of the industry.

- If the contractor's firm was recently established, it may be difficult to get a bank to respond to its initial financing needs. Often, the best starting point is the bank that has handled the *personal account* of the owner or key officers. This personal contact may be what is needed to get the contractor's "foot in the door."

- The size of the bank has an impact on its ability to service particular types of loans. It was noted earlier that asset-based lending (particularly current assets) requires a significant amount of bookkeeping, and, therefore, a reasonable sized bank staff. Many of the smaller institutions are unable or unwilling to devote resources to these areas.

All of these factors have a direct impact on the terms included in the bank note. The bank note, a legal instrument witnessing indebtedness, specifies:

- the amount of the debt obligation, or the potential limit on the debt obligation, such as with a line of credit.
- the interest rate to be charged on the debt obligation.
- the repayment terms.
- any collateral used to secure the loan.
- any other terms and/or conditions included in the debt agreement.

All of these variables are negotiable, and should be discussed at length with those lenders most suited to the needs of the firm. In many cases, just asking and stating a reasonable case will afford the contractor more favorable terms. As a firm becomes financially stronger, its position in such negotiations also becomes stronger. Lenders are very interested in establishing and maintaining good relationships with well managed and financially stable firms, and will make every effort to accommodate their needs.

The most common loan covenants deal with the contractor's maintenance of certain *financial ratios* and the establishment of other *debt relationships*. For example, the lending institution might include a ceiling on the borrower's debt ratio, a particular minimum level for the interest coverage ratio, and a minimum level for the current ratio and working capital. Lending institutions often believe that adherence to conservative guidelines will enhance the quality of the borrower's financial structure and, thus, its ability to repay the debt. If the borrower is unable to maintain the ratios specified in the indenture agreement, it may be in default of the loan agreement. If the borrower is in default, the bank may terminate the agreement and demand payment. It is certainly not in the interest of the borrower or the lender to reach this point. The lending institution prefers that the borrower successfully comply with the terms of the loan, thereby developing a strong banking relationship.

When the borrower is aware of weaknesses in the maintenance of particular ratios or other loan covenants, it is best to develop a plan to correct the deficiency and to present this plan to the lending institution. Lending institutions do not like surprises unless they are pleasant ones; the contractor should therefore face the problem up-front and work with the lender to correct the deficiency.

Negotiating with Lenders

An example of the selection and negotiation process is as follows. A contracting firm that experienced rapid growth over the first three years of operation used debt financing very prudently. The firm then committed to begin work on two major contracts, requiring a 40% increase in the size of the firm. This rapid increase could not be financed with internally-generated funds. Therefore, contact was made with the contractor's commercial bank to request an expansion of the existing line of credit by four times its previous amount. This was a true *line of credit* need, since the cash requirement was for the initial investment in labor and materials to get the jobs started. The peak borrowing need would be reached in three months, and by the end of seven months, the line of credit would be paid down to the company's normal operating level. The past banking relationship was very strong, with all payments being made in a timely manner. The contractor's profitability and gross margin levels were very positive and a strong financing package was developed (or so the contractor thought). At this point, loan discussions commenced. After initial positive indications, the bank turned down the request, reasoning that this was too much growth, too fast, and that the company did not have enough collateral to support the loan.

The situation just described has probably been experienced by most contractors at one time or another. Financing is always a double-edged sword. When it is our money that is being loaned, we want the lending institution to be very prudent. When we want a loan, we want the institution to give us a break. In the previous example, the contractor was in need of short term financing. The lending institution which had served the contractor's needs since the start of the company was no longer willing, or perhaps able, to meet the needs of the firm. The perception of the lending institution was that the contractor's financial position did not warrant granting the funds requested by the contractor. Ironically, when the contractor started operations three years ago, this was the only lending institution willing to finance any amount of start-up capital. At that time, the larger institutions had expressed no interest in establishing a relationship.

The contractor's growth simply outpaced the ability and willingness of its original lending institution to service the account. However, this was not because of the dollar amounts involved, which were relatively small. The credit line was not extended because the new needs of the contractor were such that the lending institution would have to accept a higher level of risk and assume the added responsibilities of monitoring the security that the contractor had available—primarily, *accounts receivable* and *work-in-process*.

Once informed of the bank's rejection, the contractor proceeded to present the financing application to two larger lending institutions. The contractor's relationship with the original bank remained strong, however, for each party simply realized that it was time to move on. Maintaining this good relationship is important because at some point in the future the first institution may serve another need for the contractor.

The loan request was approved by each of the two larger institutions approached. The contractor selected one over the other, based on the following criteria.

1. Most importantly, the selected institution seemed to have a better understanding of the nature and needs of the construction industry. The contractor became confident that this lending institution would respond to the firm's future changes.
2. The selected institution offered a slightly higher line of credit. This institution required additional financial reporting and paperwork, but the benefits justified the additional workload.
3. The selected institution offered a variable interest rate that was initially one half percent lower. The interest rate was tied to the bank's prime rate.

The same lending institution selected by the contractor for this expansion loan had wanted nothing to do with the financing when the company started up three years earlier. Each lending institution has its own particular "niche". The task facing the contractor is to identify the institution with the appropriate "niche" for its needs and attempt to begin a relationship with that institution. Perseverance is very important; the contractor must be willing to approach several lending institutions to find the right one.

The Loan Application: When approaching a lending institution or any other entity for financing, the contractor must be able to strongly support the request with background information. The loan application may be received by an individual, but the final decision is made by a committee. Therefore, the contractor should include all relevant supportive information when submitting the application. The contractor should prepare, in advance of the initial meeting, a package of such materials in order to expedite the decision-making process. The package should include the following materials for presentation to the financial institution.

1. Copies of the past three years' **tax returns**.
2. Copies of the past three years' **financial statements.** Statements prepared by a CPA are preferred. A reviewed statement is preferred over a compilation statement; an audited statement is better still. These three types of statements require specific services by a CPA. The audited statement is the most expensive, but carries the most weight. Lending institutions generally require at least a reviewed statement. If only in-house statements are available, the lending institution may require that a CPA examine the statements. The submission of a consolidated financial analysis (as shown in Figures 3.1–3.3) is highly desirable. Such a document reveals all ratio and trend data for a three-year period.
3. A **schedule of work-in-process, major jobs recently completed, and jobs awarded but not yet started.** Names, addresses, and telephone numbers of key references should be included. The schedules should detail the contract value, the cost incurred to date, the expected cost to complete, and expected job profit. Such information may be presented in the form of a job status report (as shown in Figure 2.9).
4. If not part of the financial statement, a **schedule of any other debt financing** used by the firm. This schedule should include the name of the lending institution, the amount of the loan, the repayment terms, and collateral.
5. An **aged schedule of accounts receivable**.
6. A **list of all major equipment and other capital assets.** The list should specify whether or not the equipment is currently serving as collateral for another loan.
7. A **list of major suppliers** and key associates that can be used as a credit reference.
8. A **cash flow forecast** for one year. The cash flow forecast and budgeted statements should point out how the debt financing will be repaid.
9. The **annual operating budget** for the coming year.
10. A **business plan,** as shown in Appendix B. This document should include the pro forma financial statements that reflect all anticipated conditions during the forthcoming year, including the new financing. The plan should be presented in the following format.
    a. A **cover page** providing a brief narrative on the history and progress of the firm. The narrative should end with the current direction of the firm.
    b. A description of **exactly what the contractor wants from the lending institution** (how much, what time period, etc.) and what the contractor is planning to do with the amount borrowed. Be specific in describing how the borrowed funds will meet the needs of the organization. The lending institution is as concerned about giving out *enough* money as it is about giving out *too much*.
11. The **marketing plan**, defining the contractor's marketplace, specific opportunities, and how the firm will acquire new projects. This plan should include a breakdown of the marketing budget and any corporate brochures used in marketing the firm.

The package should be presented in a well organized format. Not only does it contain important pieces of information, but it represents the firm. If the proposal is not presentable, the institution may never even take the time to look at it. Furthermore, proper development and packaging of the information goes a long way towards facilitating the financing process. The numbers must still speak for themselves, but a good presentation ensures that they are fairly considered.

In most cases, financing is lined up "in the nick of time." The contractor must have an answer as soon as possible to determine if he should pursue financing elsewhere. Therefore, the contractor, in presenting the application to the lending institution, should express a desire to receive a counter-offer if the proposed financing arrangement is unacceptable. The contractor should determine whether the lending institution is going to be willing to work with the firm, and if so, to what degree.

Leasing

Another way in which to acquire the use of capital assets is leasing. In a leasing arrangement, the contractor uses an asset owned by another entity in return for a periodic payment. Some prominent reasons for leasing equipment, rather than purchasing it, are listed below:

- A lease provides a way for the contractor to *conserve cash*, since the downpayments on equipment purchases may run as high as 25% to 30% of the equipment cost. A lease can usually be entered into for less than 10% of the purchase price, or perhaps nothing.
- Leasing may afford the opportunity to *accelerate tax deductions*, since the full amount of the lease payment is likely to be deductible, while only the interest portion of the debt repayment is deductible. The depreciation rate of the equipment is used to determine whether leasing or purchasing provides a better tax deduction.
- In certain types of leases, the lessee is relieved of the burden of maintaining the equipment. In the event of a breakdown, another piece of leased equipment may be available with minimum "down time" and no additional cost.
- A lease can generally be entered into for a very specific period of time. The contractor can lease pieces of equipment only during the periods of need. In this way, the commitment of funds to idle assets can be minimized.
- For very specialized pieces of equipment subject to rapid obsolescence, the contractor can avoid the risk of purchasing and then being burdened with old technology.

Certain types of assets are more likely to be leased than others. For example, contractors often lease such capital assets as office space, warehouse facilities, and equipment yards. For equipment that is very specialized, subject to rapid obsolescence, used infrequently, or just impractical to buy, leasing may also be a practical and popular solution.

Leases can be structured in a variety of formats. The two major types are *operating leases* and *capital leases*. Each has different applications and affects the firm's financial structure in different ways.

Types of Leases

Operating Leases: An operating lease allows the contractor to use an asset (such as a bulldozer) for a limited period of time for a fixed periodic payment. The lease may specify that the lessee (the party leasing the property from the owner) maintain the equipment in normal operating condition, *or* that the lessor (the owner of the property) maintain the equipment.

An operating lease does not provide for a change of ownership or the payment of a "bargain" purchase price at the end of the lease term. The term of the lease is generally less than 75% of the useful life of the property, and the present value of the lease payments is less than 90% of the fair market value of the property. Accounting principles provide for very specific rules in the treatment of leases on the contractor's books.

One of the advantages of an operating lease is that it provides *off balance sheet financing*. This means that the contractor has, in essence, a debt obligation that does not appear on the firm's balance sheet. By not appearing on the balance sheet, this debt does not affect the contractor's financial ratios (especially debt ratios). This approach may be necessary if the contractor is on the "borderline" with restrictive covenants in other debt agreements, and will improve the firm's bonding capacity.

Capital Leases: A capital lease has at least one of the following criteria:
1. The lease term exceeds 75% of the useful life of the asset.
2. The present value of the lease payments exceeds 90% of the fair market value of the asset.
3. The lease provides for a transfer of ownership or a "bargain" purchase at the end of the lease term. Unlike an operating lease, the obligations of a capital lease must be reflected in the firm's balance sheet—either as a line item under *Liabilities* or, possibly, in the notes attached to financial statements.

Evaluating the Lease

While a lease arrangement can be very advantageous to a construction firm, it must be entered into just as cautiously as any other contractual obligation. A number of lease terms will directly influence this decision. It is advised that an attorney and/or CPA be consulted before executing any lease agreement. In addition to a financial analysis of the lease versus purchase decision, other factors to consider and negotiate in a leasing arrangement include:

- Periodic payments, interest rates, and penalties for late payment.
- Number of years lease is to run and penalty for cancellation.
- Condition in which equipment is to be returned to the lessor.
- Options for renewal and associated terms.
- Responsibility for minor and major maintenance.
- Value at which equipment may be purchased at any time during the lease or upon termination.
- Ability of the firm to sublease the equipment if so desired.
- Responsibility for insurance.

To evaluate the actual cost of the lease, the contractor should calculate the present value of the after-tax cost of all future payments associated with the lease. This calculation can be made using the discounted cash flow procedure discussed in Chapter 6.

For example, suppose our example contracting firm, Eastway Contractors, is interested in leasing a piece of equipment for four years. The annual lease amount is $24,000, and at the end of the lease term the equipment can be purchased for $20,000. Eastway will probably purchase the equipment at that time. A cost of capital of 12% will be used to discount the future cash flows, since that is the interest rate that would be charged by the lender if Eastway purchases the equipment. If Eastway is in the 30% tax bracket, the after-tax cost of the lease will be $16,800. The present value of an annuity of $16,800 for four periods at 12% is $51,027 ($16,800 × 3.0373, Figure 6.7). (The present value factor assumes that the lease payments are annual and occur at the end of the year.) While this is not always an exact estimation, the exercise should provide a reasonable comparison of the relative cost of financing.

The next cash outflow is the purchase of the equipment at the end of the lease term. The purchase price of $20,000, multiplied by the present value interest factor for a single cash flow occurring at the end of the fourth year at 12%, .6355, yields a present value outflow of $12,710. The present value of the cash outflows associated with leasing can be summarized as follows:

Present value of the after-tax lease payments	=	$51,027
Present value of the purchase of the equipment	=	$12,710
Total present value of cash outflows		**$63,737**

In this example, it is assumed that the operating cost of the asset is the same whether the asset is owned or leased. The present value analysis of the cash outflows associated with leasing must be compared with the present value analysis of the cash outflows associated with ownership, which is covered in detail in Chapter 6 on Asset Management. In certain instances, the benefits of leasing clearly outweigh the benefits of ownership.

Summary

Financing a firm's assets is an important management function. To determine the best methods for financing, the long term development of the firm and the role that debt will play in promoting growth and profitability should be analyzed. The maturities of the debt obligations should be matched with the timing of returns generated by the underlying assets.

Short term financing is generally more risky than long term financing, since there is a more frequent risk of rising interest rates. Long term financing, while generally more expensive, allows for more accurate forecasting of cash flow requirements.

The effective annual cost of each loan option should be considered. Features such as loan discounting, compensating balances, and loan placement fees have the effect of increasing the cost of financing above the stated interest rate.

When applying for financing, the contractor should assemble a package presenting the firm in a favorable way. This package should include historical information about the firm's financial performance, projected future performance, professional references, major creditors, and requested financing arrangements.

Leasing, if appropriate, provides an alternative financing vehicle to the standard purchase. Some of the advantages of leasing include a smaller cash requirement, reduced risk of equipment obsolescence, and a stronger repair and maintenance routine.

Chapter Eight

TAX MANAGEMENT

Chapter Eight

TAX MANAGEMENT

During the last few years, Congress has passed a major tax bill almost every year. The result is a tax system with untold complications, difficult to interpret and apply, even for the most experienced CPA's. The Tax Reform Act of 1986 made sweeping changes in the way that businesses, especially contractors, are taxed. Construction firms must deal not only with the maze of federal, state, and local rules that affect all businesses, but also with an additional set of overlapping rules, in order to compute its tax burden.

Tax management should play a major role in the development of every contractor's business plan—at the very outset. Today, there are three major factors that influence how much tax contractors will pay and how much control they will have in minimizing their tax burden. The first factor is the *form* (such as a corporation or a partnership) of the business organization. Decisions regarding the form of a business organization will affect the timing and amount of taxes paid by the business.

The second factor is the *volume* of business. Volume has become particularly important as a result of the 1986 Tax Reform Act. The Act made sweeping changes in the way businesses, especially contractors, are taxed. The methods of accounting and tax deferrals available to a contractor vary dramatically, depending on the volume of business, since the tax rules change at different revenue volumes.

The third factor under the contractor's control is his firm's *method of recognizing income* for income tax purposes. This chapter addresses these and other complex tax considerations that affect construction firms.

The one lesson that every experienced contractor has learned is that good tax management means using all the available tax deferrals and tax avoidance mechanisms. Maximizing these mechanisms can mean the difference between business success and failure. Contractors have historically enjoyed certain tax treatment benefits that allowed them to defer substantial portions of their tax liabilities year after year. If a contractor plans for maximum use of tax deferrals and tax avoidance, he is essentially borrowing money from the government, interest-free, for an unlimited term, and with no collateral. A word of caution, however, about tax deferral: it is exactly that, a *deferral*. It is not elimination. Building tax deferrals creates interest-free capital, but without proper monitoring of tax planning, these same tax deferrals can become overnight liabilities, with the government demanding payment. Proper planning should include the timing of tax payments, so they will not create a hardship on the cash flow of the firm.

The tax planning process is ongoing. It starts with the initial organization and selection of accounting methods, continues with the monitoring of income throughout each year, through to follow-up at year's end to ensure proper implementation of the plan.

Given today's complex tax laws, it would be impossible to explain all of the nuances and intricacies of tax management for contractors in one "readable" chapter. The information in this chapter should be used as a general guide to those aspects of tax law that should be considered by every contractor in tax management and planning. These rules should be implemented with professional advice from a CPA or attorney who has experience working with construction contractors.

Organizational Structure

The legal form of organization chosen for a firm has a major impact on its taxation and should be a primary consideration to a new business. There are three basic forms of legal organization:

1. Sole Proprietorship
2. Partnership
3. Corporation
   a. Regular "C" Corporation taxation status
   b. "S" Corporation taxation status

All construction companies must deal with risk. The legal liability associated with doing business as a sole proprietorship or partnership often leaves businesses no choice but to organize as a corporation, in order to protect the personal assets of the owner. This chapter, however, will concentrate only on the *tax consequences* of operating under these various forms of business. Coverage of the legal and other considerations of these organizational forms is beyond the scope of this discussion.

Sole Proprietorship

Many small contractors start their businesses as sole proprietorships because this is the least costly and simplest way to operate. The sole proprietorship is not taxed as a separate entity. All of the income (or loss) flows directly through to the owner on Schedule C of Federal Form 1040, and is taxed directly to the owner. Other than simplicity, the major benefits of this form are the owner's ability to deduct start-up operating losses and, once profitable, lower individual tax rates.

While the sole proprietorship form may be suitable for small, start-up companies, it is less practical when used by a growing or large contractor. Used for growing companies, sole proprietorship may raise financial reporting problems. Furthermore, it prevents the owner from utilizing the tax deferral mechanisms and employee fringe benefits that are available through a corporation.

Partnership

The partnership form of organization is most often used when two or more people decide to join their capital and other resources to operate a business. For tax purposes, the partnership is treated in much the same manner as a sole proprietorship. It differs in that it is considered a separate entity, thereby eliminating some of the financial reporting problems associated with a sole proprietorship. Taxable income or loss is passed through to the individual partners and is reported in their individual tax returns.

One of the benefits of operating as a partnership, aside from being recognized as a separate business entity, is the ability to control distributions of earnings and income to the partners (owners). The tax rules for allocations of income and cash distributions are very complex. The advice of professional tax counsel is recommended for any arrangement other than a pro rata allocation of all items between partners. While the control of allocations is normally considered a benefit, it may become a detriment when there is insufficient cash flow or the company is losing money. Other potential tax difficulties in a partnership arrangement may be encountered during the liquidation of a business, or when one partner wishes to withdraw from the business. Both cases are more complicated and involve associated tax issues beyond simply selling corporate stock.

Corporations

Because of the limitation on personal liability, as well as administrative convenience and tax planning opportunities, the corporation is the most common form of organization for construction firms of any substantial size. The corporation is a separate legal entity. As a result, its ownership is more easily transferable. Corporations can conduct business with or without the direct involvement of the owners. These are good business reasons for operating in the corporate structure. However, there are tax considerations that may be advantageous or disadvantageous, depending on the particular situation of the stockholders. The following paragraphs describe some of the tax considerations of operating as a corporation.

Tax Rates

As a result of the 1986 Tax Reform Act, the rate for taxable income in excess of $75,000 for *corporations* is now higher than the tax rate for *individuals* above that amount. Prior to this change, corporations enjoyed a lower tax rate than individuals at all income levels. Figure 8.1 illustrates the tax rates at various income levels for corporations and individuals. The highest marginal tax rate for corporations is now 39%, versus 33% for individuals; for incomes in excess of $335,000 the rate for a corporation is 34%, versus 28% for an individual.

What is the tax *benefit* of operating as a corporation? As illustrated in the tax table, corporations benefit from lower tax rates on income *below $75,000*. A corporation that can keep its yearly taxable income below this amount can accumulate more "after-tax" capital than an individual. For this reason, it may be advantageous to operate several small firms, each dealing with its own specialty or geographical area. Construction companies require large amounts of working capital for such things as purchasing equipment or funding retentions receivable. The lower tax rates of a corporate structure allow the smaller, growing contractor to accumulate this working capital.

Taxable Years

A newly formed regular corporation may choose any month for the end of its taxable year. This flexibility allows the corporation to choose the most beneficial time in its tax planning. For example, a contracting company performing the bulk of its work during the summer may want to end its tax year in the late winter when cash and taxable profits are at their lowest. The use of a fiscal year (rather than a calendar year) also allows some flexibility in the timing of payments to both officers and shareholders. A regular corporation with a February year's end might pay bonuses to its owners in January and February in order to reduce its taxable income. The individuals would be taxed during the calendar year, thus allowing for a possible deferral of the tax payments until sometime later in the calendar year, or possibly even into the next calendar year.

Comparison of Tax Rates Applied to $100,000 of Taxable Income for Corporations and Individuals

Taxable Income	Corporate Income Tax		Individual Income Tax**	
	Rate	Amount	Rate	Amount
$0–50,000	15%	$7,500	—	—
$50,001–75,000	25%	6,250	—	—
$75,001–100,000	34%	8,500	—	—
$0–29,750	—	—	15%	$4,463
$29,751–100,000	—	—	28%	19,670*
Federal Income Tax		$22,250		$24,133

*Does not include the effect of the surtax described below.
**Married, filing jointly.

For both corporations and individuals there is a tax surcharge on incomes in excess of $100,000. For corporations with income in excess of $100,000, the tax is increased by 5% of the amount over $100,000 up to a maximum of $11,750 in additional tax. This surtax eliminates the tax savings of the lower tax rates (15% and 25%) for corporations with taxable income over $335,000.

For individuals, there is also a 5% surtax. For example, for married individuals filing jointly, the 5% additional tax (for incomes over $71,900 up to $149,250) effectively eliminates the tax rate differential of the 15% tax rate bracket when taxable income is over $149,250.

Figure 8.1

Retirement and Fringe Benefit Plans

Since a corporation usually employs its owners, they are generally eligible for the same nontaxable fringe benefits as other employees. Regular fringe benefits include health insurance, life insurance, disability insurance, deferred compensation, pension and profit sharing plans, cafeteria plans, and other similar arrangements. Many of these benefits would be taxable under the other forms of organization. It is important to remember that the IRS has strict rules of nondiscrimination regarding benefits for the officers/owners and those provided to other employees. These rules must be met to qualify as nontaxable fringe benefits for the owners.

Corporate Liquidations and Double Taxation

The Tax Reform Act of 1986 substantially changed the taxation of corporations that are sold and liquidated. Prior to the change, the shareholders of a corporation could sell the corporation's assets, liquidate the corporation, and only be taxed once, at the individual level. The Tax Reform Act of 1986 repealed what was known as the "General Utilities Doctrine," which allowed such liquidations with only a single level of taxation. This development may be of little importance to a new business owner with no plans for selling or liquidating his business. However, it does affect the economies of buying and selling businesses that are operated in corporate form. The effect is that both the shareholders and the corporation pay tax on the basis of the fair market value of the corporation's assets, when it is liquidated. As a result, sellers will want to sell the stock of their companies (rather than the assets) and buyers will be unable to increase the basis of the corporations' assets up to the price paid, if all they buy is the stock of the corporation. There are some exceptions to these rules for businesses with values of less than $5 to 10 million through January 1, 1989.

Passive Losses

Regular corporations enjoy an additional benefit not available to individuals as a result of the Tax Reform Act of 1986. A closely held regular corporation may deduct from its active trade or business income those losses that relate to passive investments. Passive investments are broadly defined and include any business activity in which the taxpayer does not "materially" participate in the activity. The most common source of passive loss is from rental real estate activities, such as operating office or apartment buildings. Generally, a limited partnership interest is also considered passive.

The regular corporation's ability to utilize these passive losses and passive tax credits has both a benefit and a possible cost. The primary benefit is that the passive losses available to the corporations, (which can result from depreciation of real property and thereby utilize little or no cash) allow them to shelter current business income in the corporation and to accumulate pre-tax capital more quickly. However, there are two pitfalls:

- The Tax Reform Act of 1986 modified the corporate alternative minimum so that taxes apply to more corporations, reducing or even eliminating the tax savings. The use of passive losses to reduce current taxable income may increase the likelihood of tax applied to a regular corporation. (The corporate alternative minimum tax is discussed later in this chapter.)

- If the asset generating the losses (such as an office building) is owned by the corporation and appreciates in value, the corporation may be subject to double taxation when the corporation sells the asset or liquidates. In other words, the interim savings gained by holding such assets in a corporation may be more than offset when it is sold.

Using passive losses to shelter corporate income should be a consideration for corporations that are not planning to sell their businesses in the foreseeable future and need to accumulate capital within the corporation to operate or expand their businesses.

Unreasonable Compensation and Accumulated Earnings Tax

Over the years, the IRS has attacked closely held regular corporations in a number of ways, two of the most common being *unreasonable compensation to owners* and the *accumulation of excess earnings in the business*. If the contractor pays out large salaries to owners, the IRS may claim that these payments are excessive (for the services rendered) and are merely disguised dividends to the shareholder. If the IRS determines that a portion of the salary represents dividends, this portion is nondeductible to the corporation. The result is a form of double taxation (tax on the income to the corporation and tax on the income to the shareholder).

To strengthen the owner/employee case against such a claim by the IRS, the corporation should have employment agreements with all of its key employees. Such agreements should include the specific terms of employment, services to be rendered and, most importantly, a formula to support the amount of compensation. Any bonuses should be based on a predetermined formula, which might be outlined in the employment agreement or corporate minutes.

Regular corporations may also encounter the IRS's assertion that the corporation is retaining earnings beyond what is required to conduct its business. The outcome can be the application of the accumulated earnings tax, which is a penalty for the corporation's failure to pay dividends.

A successful argument against these IRS assertions is that construction companies always require large capitalization and working capital balances. Sureties require construction companies, as part of their indemnity agreements, to maintain certain equity and working capital levels as a condition to the issuance of bonding.

"S" Corporations

The "S" corporation mixes the non-tax benefits of the corporate form of business with the tax benefits of operating as a sole proprietorship or partnership. Generally, the "S" corporation format eliminates the tax paid by the corporation by assigning the corporation's taxable income directly to the shareholders.

Up until the Tax Reform Act of 1986, many contractors avoided the use of "S" corporations because the surety industry did not look favorably upon them. Sureties were often uncertain of the actual tax liability of the shareholder. They were concerned that the shareholders would withdraw large amounts of capital to satisfy personal tax liability obligations (due as a result of the corporation's income on their individual tax returns). For this reason, many sureties penalized "S" corporations in their calculation of maximum allowable bonding limits—beyond what would have been the case if they operated as a regular corporation. Contractors with limited bonding capacity were disinclined to consider "S" corporations as an alternative to a regular corporation status.

The Tax Reform Act of 1986 substantially changed the benefits and rules of operating as an "S" corporation. The changes relate primarily to the use of fiscal years, the election by existing corporations, and the methods of income recognition. Since the "S" corporation form now offers contractors numerous business and tax benefits, most sureties have been forced to deal with the "S" corporation issue.

The basic rules for a firm to qualify as an "S" corporation are relatively simple. The law provides that:

1. An election must be filed with the IRS within 75 days after the beginning of a corporation's fiscal year in order to be effective for that year.
2. There can be no more than 35 shareholders.
3. The corporation cannot own another corporation as a subsidiary.
4. The corporation cannot have as a shareholder a nonresident alien, a corporation, or certain entities other than individuals.

Features of "S" Corporation Status

One benefit of operating as an "S" corporation is that tax rates are lower for individuals than they are for regular corporations. For corporations with large earnings, the differential between the 28% tax rate for individuals and the 34% tax rate for regular corporations could be substantial. Other factors are:

- The double taxation problem that may affect regular corporations does not exist for "S" corporations. Usually, "S" corporations can sell assets and the resulting income is taxed only once, to the shareholder. There are, however, specific rules that limit this benefit for newly-electing "S" corporations. These rules provide for taxation at a corporate level on any gain (such as appreciated property) that existed before the conversion to an "S" corporation and which is recognized by the "S" corporation within 10 years after the date of its election. Certain closely held corporations are exempt from these rules through January 1, 1989. Before electing "S" corporation status, existing corporations should thoroughly investigate the possibilities for double taxation.
- A net operating loss of a regular "C" corporation at the time it elects to be an "S" corporation remains with the corporation and is only available to offset income on built-in gains or income as a regular corporation, if the election is revoked in the future.
- Generally, "S" corporations must end their year in December. The law provides a special exception (which must be applied for by newly electing "S" corporations) which allows them to use a September, October, or November year end. Any other month can be used only if the corporation can show a compelling business purpose to the IRS, which is unusual.
- The shareholder's deduction for the net operating loss of an "S" corporation is limited to the shareholder's adjusted basis of the stock and any debt that the corporation owes him. The nondeductible portion of the loss is carried forward to future years, and will be deductible when there is enough tax basis to allow it.

- The allocation of all items of income, losses, and credit are in proportion to the shareholder's stock interest. There can be no "special allocations" of items, as can be allowed in a partnership.
- In some circumstances, a separate corporate level tax can be imposed on passive income (rents, royalties, interest) in excess of 25% of total income. The limitation on passive loss items applies to the shareholders as individuals, not to the corporation.
- An "S" corporation is *not* subject to limitations on the use of the cash basis method of reporting income, which will be described later. This can be a significant benefit to contractors. One word of caution: a newly-electing "S" corporation previously using the cash basis may be subject to corporate tax on its unrecognized receivables. This could be very costly for the contractor if it elects to report on the accrual basis, *percentage of completion* basis, or *completed contract* basis.
- The "S" corporation is not subject to the book-tax difference add-back provisions of the corporate alternative minimum tax. As a result, significant tax savings may be possible for small contractors using the completed contract or cash basis methods of income recognition.

Any existing corporations that may be considering the possibility of electing "S" corporation status should evaluate and plan carefully to ensure that the election will not cause the corporation an immediate tax liability.

Methods of Income Recognition

There is probably a wider choice of income recognition methods for construction contracting than for any other industry. The Tax Reform Act of 1986 and Deficit Reduction Act of 1987 further complicate the choice of income recognition methods. Small contracting companies with average gross receipts of less than $10 million and individual contracts of less than two years' duration are allowed exceptions from some of the more cumbersome rules and may use the completed contract method of accounting without restriction. Larger contractors must abide by new accounting rules under these two new tax acts.

Not only are there a number of alternative methods of income recognition available to contractors, but changing methods after filing one's initial tax return is very difficult. Consequently, it is important that contractors review their business plans to take maximum advantage of the income recognition alternatives at the very outset of business. There are four basic methods of income recognition available to contractors:

1. Cash Method
2. Accrual Method
3. Completed Contract Method
4. Percentage of Completion Method

These methods can be combined at the option of the contractor. Alternatively, the contractor may be required to use the newly created *percentage of completion-capitalized cost method*.

Cash Method

Using the cash method, income consists of all revenue actually received or constructively received (cash receipts) for work and expenses. Expenses include amounts actually paid for costs, such as labor, materials, subcontractors, and operating expenses. Most contractors find this the easiest method to use because it does not include any complicated cost estimates and the amounts of income and expenses are easily determined.

The major advantage of the cash method is that it gives the contractor the opportunity to manipulate the amount of taxable income recognized by acceleration or deferral of either income or expenses. Contractors often overlook this method. However, it should be strongly considered as it has the potential for creating substantial tax deferrals.

The Tax Reform Act of 1986 prohibits the use of the cash method by regular "C" corporations with average annual gross receipts of $5 million or more based on an average of the three prior years. An "S" corporation, on the other hand, is not restricted in its use of the cash method and, as previously discussed, offers many of the same benefits of a regular corporation. The major disadvantage of the cash method for financial reporting is that it is not accepted by lenders or sureties for any contractor of substantial size. Contractors wishing to use the cash method would also have to maintain an additional set of financial records, using a method that these parties would accept.

Accrual Method

The accrual method of accounting is less flexible for tax planning purposes than the cash method, but it is far more accurate for financial reporting purposes. Many contractors, therefore, maintain their records in an accrual format. Under the accrual method, income includes all amounts that the contractor is entitled to receive when the amount can be determined and all events have occurred to fix the *right to receive the income*. Expenses include liabilities incurred that obligate the firm to payment of a determinable amount with reasonable accuracy. The accrual method thereby provides a more complete picture of the firm's true financial performance.

Figure 8.2 is a table showing the calculation of taxable income using the cash and accrual methods. The accrual method is not commonly used by contractors for tax reporting purposes for several reasons. First, it lacks flexibility. Second, contractors often have higher balances of receivables than payables. Using the accrual method, the result is payment of taxes on income not yet received. Further, if the contractor front end loads a contract, billing substantially more than he has to pay out in the early stages of a job, he may be taxed on profits that are inflated by the overbilling.

One principal advantage of the accrual method has often been overlooked by contractors. Several court decisions support the fact that a contractor does not have to recognize the income from retentions until he has the right to receive the payment from such retentions. Using this deferral method can substantially increase the amount of deferred income for a contractor. A contractor who has historically included retention income, as many have done, must apply to the IRS for a change in accounting method if they wish to defer the recognition of income from retentions.

If a contractor holds an inventory of finished or in-production items for resale in the normal course of business (for example, a contractor who operates as a building supply store), he must use the accrual method of accounting for that portion of his business. The law allows the contractor in this case to utilize a "hybrid" method that may include the accrual method for the retail/wholesale portion of his business and the cash method for the construction contractor portion.

Long Term Contract Methods

The IRS permits contractors to utilize several additional methods for recognizing income from long term contracts. Long term contracts are specifically defined by the IRS as *those contracts in which the work spans more than one fiscal year.* It should be noted that a contract may require less than 12 months to complete, but still span two fiscal years and as such be classified as a *long term contract.*

Calculation of Taxable Income Using the Cash vs. the Accrual Method		
	Cash Method	Accrual Method
Total Billings on Contract for the fiscal year	$1,000,000	$1,000,000
Less: Retentions Receivable at 10% of total billings	(100,000)	(100,000)
Contract receivables from prior month not yet received at year-end	(100,000)	
	800,000	900,000
Gross income recognized (1)		
Total Direct Costs and G&A Expenses incurred	700,000	700,000
Less: Costs and expenses not paid at year-end	(50,000)	
Total Direct Costs and G&A Expenses (2)	650,000	700,000
Taxable Income (1–2)	150,000	200,000

Figure 8.2

Under prior law, there were generally two methods used to recognize income for long term contracts. The contractor could choose between these methods, but had to use the same method for all long term contracts. These methods were:

- **Percentage of Completion Method**—whereby the income on each individual contract is recognized proportionately as the work is completed.
- **Completed Contract Method**—whereby the full amount of income and expenses for each contract is deferred until the contract is completed. Upon completion, *all* of the income and expenses are recognized.

When planning, it is important to remember that the long term contract rules apply on a contract-by-contract basis, not to the contracting company as a whole. Therefore, as rules change, they must be applied to each contract individually. Applying the rules in this manner makes the administration and accounting for them more time consuming—both in terms of record keeping and for planning purposes.

As a result of the Tax Reform Act of 1986, contracts entered into after March 1, 1986 must use either the *modified percentage of completion method* or the *percentage of completion-capitalized cost method*. Figure 8.3 is a table outlining the various options and requirements.

The table in Figure 8.3 indicates, by the size of the contractor, those methods available, required, or prohibited. Although some of the methods are available to contractors of a certain size (such as the 70–30 method for a contractor with less than 5 million of gross revenue), they do not have any tax advantage and should not be considered.

The law has now distinguished contractors by size and contract duration in determining the methods of accounting available to them. The rules for large contractors are more cumbersome and less advantageous than those for small contractors. To clearly outline the differences in rules, they are discussed separately.

Methods of Accounting for Large Contractors

The following section presents the accounting methods generally used by large contracting firms. "Large contractors" are defined as those earning more than $10,000,000 in annual gross receipts, or having contracts of more than two years' duration.

Modified Percentage of Completion

The percentage of completion method was modified in two ways for large contractors by the Tax Reform Act of 1986. First, the calculation of percentage of completion was restricted to the "cost-to-cost" method. This method determines the percentage of completion by comparing the costs of a contract incurred by the close of the tax year to the total estimated cost of the contract.

Income Recognized in Current Year =

$$\frac{\text{Cost incurred to date}}{\text{Total estimated cost}} \times \text{Contract Price} - \text{Income recognized in previous years on contract}$$

Under prior law, the percentage of physical work completed could be used to determine the percentage of completion. This method is still available to contractors with under $10 million in annual gross receipts, using the regular *percentage of completion* method. The second major modification is the application of the uniform cost capitalization rules to all contractors not eligible for the small contractor exemption. Generally, these rules require all costs that benefit (or are incurred because of) a long term contract to be allocated to the contract. The specific rules for this situation are discussed in more detail later in this chapter.

Comparison of Income Recognition Methods by Size of Contractor Volume

| | Contractor Size | | |
| | Small Contractor Exemption | | |
Method of Accounting	Contractors with less than 5 million of Gross Revenue	Contractors with 5 to 10 million of Gross Revenue	Contractors with greater than 10 million of Gross Revenue
Cash Basis Method	Available as optional method	Unavailable	Unavailable
Accrual Basis Method	Available as optional method	Available as optional method	Available in conjunction with either long term contract methods
Modified Percentage of Completion Method (with Cost Capitalization rules)	Not required	Not required	Required Option
Regular Percentage of Completion Method (without Cost Capitalization rules)	Available as optional method	Available as optional method	Unavailable
Completed Contract Method	Available as optional method	Available as optional method	Unavailable
Percentage of Completion-Capitalized Cost Method ("70–30 method")	Not Required	Not Required	Required Option

Figure 8.3

Percentage of Completion Capitalized Cost Method

This is an entirely new method of income recognition introduced by the Tax Reform Act of 1986, and modified in 1987. It applies only to contractors with more than $10 million in average gross receipts for the preceding three years or for contracts of more than two years' duration. For contracts entered into after February 28, 1986, but before October 14, 1987, the contractor must report 60% of the income from the contract through the firm's regular method of accounting (such as the completed contract method, cash method, or accrual method) and the remaining 40% through the new, modified percentage of completion method. For contracts entered into after October 14, 1987, the law requires 70% of the income to be recognized on the basis of modified percentage of completion and 30% through the regular method.

The "Look Back" Rule

Contractors using either the modified percentage of completion method or the percentage of completion capitalized cost method, must perform an additional computation after each contract is completed. When a contract is completed and income has been recognized under either of the above methods (Modified Percentage of Completion or Percentage of Completion Capitalized Cost), the taxpayer is required to compare the income recognized and the amount of taxes actually paid during each year of the job to the estimated income and tax.

The *actual* contract price and costs are used to compute the overpayment (or underpayment) of tax which would have been payable for each tax year. Based on this calculation, the contractor will then owe interest to the IRS (or vice versa), on underpayments (or overpayments) of tax. There is no *additional tax* due in these cases, only *interest* or any underpayment of tax in prior years. The IRS must pay interest on the contractor's overpayment of tax. By recomputing, contractors can avoid underestimating profits.

Cost Capitalization Rules

Before the Tax Reform Act of 1986, contractors had some discretion in allocating indirect and service costs to their contracts for tax purposes. They were subject to the "extended period long term contract" regulations only if their gross receipts were over 25 million dollars. These regulations are the basis for the cost capitalization rules that apply to contractors not eligible for the small contractor exemption.

The cost capitalization rules apply to all contracts entered into after February 28, 1986. Excluded are contracts that are estimated to be completed within a two-year period and that are executed by a contractor whose average annual gross receipts for the three preceding years do not exceed $10,000,000 (the small contractor exemption). Regardless of the accounting method used, if the contractor and contract do not meet both these conditions, the cost capitalization rules must be applied to all long term contracts. The requirement to capitalize more contract-related costs means that deductible expenses are reduced and taxable income increased accordingly for contractors who are subject to these rules.

The other costs associated with cost capitalization rules are the additional administrative accounting burden and the tax planning and preparation. There are several methods allowed for the allocation of indirect costs for service and support activities. All of them carry the burden of extensive additional record keeping. The use of computerized accounting and record keeping systems becomes almost a requirement for a contractor who is subject to these rules. Many contractors may find that in order to comply, they must change their job cost accounting procedures.

Indirect Cost Allocations

To allocate indirect costs (as defined in Figure 8.4), the contractor has a choice of either directly tracing and charging the costs to individual contracts, or using a "burden" rate. The burden rates may apply these costs based on any of several factors, such as *unit costs* or *direct labor hours*.

General and Administrative Cost Allocations

General and administrative costs (outlined in Figure 8.4) are not as easy to allocate as direct and indirect costs. Research and development costs are specifically excluded by the law, as are the costs of unsuccessful bids and proposals, and marketing expenses. To reduce the amount and number of cost types to be allocated, the contractor should analyze expense types and job responsibilities for each person on the administrative staff. For every individual who can reasonably be determined to be acting in a management function (rather than contract related), the labor cost plus that individual's share of rent, depreciation, office supplies, and support salaries will also become a deductible expense rather than being capitalized.

The common costs for each department that are allocable as indirect contract support costs should be segregated into pools of indirect costs. The associated general and administrative support costs, such as rent, utilities, and supplies, should also be included. These pools of contract support cost should then be allocated to individual contracts in a consistent and logical manner similar to other indirect costs, using a burden rate factor. All the costs included in each pool must be allocated in full; there can be no under-applied indirect or general and administrative costs, such as allowable under the old law.

Additional costs are allocable to cost-plus type contracts or federal contracts. However, the costs must be identified by the contractor's internal records as part of the contract costs (pursuant to a requirement under the contract).

Interest Allocations

Interest expenses must be allocated to all contracts, even if the contractor is not required to utilize the cost capitalization rules (due to the small contract exception). Generally, interest expense incurred in connection with a long term contract must be allocated under the same rules that apply to all production activities, even those that are not long term contracts.

Production period interest not specifically identifiable with a particular asset or contract should be allocated to long term contracts using what is commonly referred to as the *avoided interest method*. This method calculates the amount of interest to be capitalized to a particular asset (long term contract) as the amount that would have been theoretically avoided (not incurred overall by the company) if the activity or asset had not been performed or constructed. This method nets all contract progress payments against the asset's contract cost. As such, for most construction contracts for which the contractor receives monthly requisitions of progress payments from the owner, the amount would be considered *de minimis* and would not have to be included in the contract costs.

The interest associated with equipment utilized in the performance of a long term contract cannot be avoided. Generally, such interest is incurred specifically for performance of work on contracts.

The period to which interest must be allocated begins on the contract commencement date. This is the date on which the contractor first incurs any costs under the contract (not including bidding and estimating costs). For an accrual basis taxpayer, this date is the later of the following dates: the date on which 5% of the total estimated costs are incurred, or the contract commencement date. The period ends on the contract completion date.

There are some exceptions to the interest allocation rules. For example, contracts of less than $1,000,000 in total cost and those that will be completed in less then one year do not require any interest allocation. Contractors using the *percentage of completion* method are exempt as the net effect would, theoretically, always be *de minimis*.

Small Contractor Exemption

The following section covers the small contractor exemption. "Small contractors" are defined as those having less than $10 million in gross receipts with contracts of less than two years duration.

Contractors coming under this exemption may utilize either the *completed contract method* or the "regular" *percentage of completion method*. They are not subject to the cost capitalization rules for any contracts regardless of their accounting method and may use the actual amount of work completed to measure the percentage of completion.

Completed Contract Method

This method of accounting is available only to small contractors. Its most important feature involves the almost indefinite deferral of substantial amounts of income. It should be noted, however, that as of this writing, it appears highly likely that Congress will soon eliminate the completed contract method of revenue recognition as an alternative available to contractors of any size. The basic benefits, along with possible disadvantages of this method as currently allowed, are:

Advantages:
- The contractor can defer 100% of the income from each long term contract until completion of the contract. The contractor can, therefore, control the timing of the recognition of income to a certain extent.
- Income can be deferred until the contract sum is known, thereby eliminating the possibility of having to recognize "phantom" income.
- The contractor can currently deduct certain costs and expenses that directly benefit the contract.
- A fiscal year (based on the geographic location and type of work the contractor does) can be used; most contracts will be completed after the close of the firm's fiscal year.
- The contractor may be able to mix construction management contracts (which are typically not eligible for long term contract income reporting) and service work with long term contracts in order to control taxable income from year to year.

Treatment of Costs and Expenses for Tax Purposes under the Cost Capitalization Rules

	Capitalization Rules for Exempt Contracts (1)	Capitalized Costs Under New Cost Capitalization Rules (2)
Direct Construction Costs		
Material	Capitalize	Capitalize
Labor	Capitalize	Capitalize
Indirect Construction Costs		
Repair and maintenance of contract-related equipment	Capitalize	Capitalize
Utilities and rent for equipment or attributable for facilities used in the performance of contracts	Capitalize	Capitalize
Indirect labor incl. related benefits and taxes	Capitalize	Capitalize
Production supervisory labor incl. related benefits and taxes	Capitalize	Capitalize
Indirect materials and supplies	Capitalize	Capitalize
Small tools and equipment	Capitalize	Capitalize
Quality control & inspection	Capitalize	Capitalize
General and Administrative Expenses		
Research & development (contract related)	Expense	Capitalize
Marketing, advertising & selling	Expense	Expense
Other distribution expenses	Expense	Capitalize
Tax depreciation in excess of financial depreciation	Expense	Capitalize
Local and Foreign Income Taxes	Expense	Expense
Past Service Costs of Pensions	Expense	Capitalize
Administrative (General)	Expense	Allocable
Officers' Salaries (General)	Expense	Allocable
Bidding—Successful	Expense	Capitalize
Bidding—Unsuccessful	Expense	Expense
Depreciation on Idle Property	Expense	Expense
Taxes such as payroll, sales and any other taxes related to facilities or equipment for contracts	Capitalize	Capitalize
Financial Statement Depreciation and Amortization	Capitalize	Capitalize
Employee Benefits	Capitalize	Capitalize
Cost of Rework Labor, Scrap and Spoilage	Expense	Capitalize
Direct Administrative Expenses	Expense	Capitalize
Officers' Salaries (Direct)	Expense	Capitalize
Insurance Cost	Expense	Capitalize
Cost of Strikes	Expense	Capitalize
Current Service Pension and Profit Sharing Costs	Expense	Capitalize

Interest (capitalizable in certain circumstances, not including situations where the contractor is receiving progress payments)

(1) Exempt contracts are those where the contractor has less than $10 million in annual gross receipts for the past three years and individual contracts of less than 2 years' duration.
(2) For nonexempt contracts entered into February 1, 1986.

Definitions of Cost Capitalization Terms

Expense—Those costs that are eligible for deduction in the period in which they are paid or incurred and do not have to be capitalized as part of a contracts cost.

Capitalize—Those costs which must be allocated to the total costs of all contracts or a particular contract, if specifically identifiable as *direct* costs.

Allocable—Those costs which are not specifically identifiable as a contract-related cost and are service- or support-related. They must be allocated between those activities which they support that are contract-related and those that are management-related. The contract-related portion must be capitalized; those that are not contract-related can be expensed currently.

Figure 8.4

Disadvantages:

- The contractor may be subject to large swings in the amount of taxable income it recognizes from one year to the next, thereby creating complications in cash flow planning.
- The difference in income recognized between the percentage of completion and completed contract method of accounting is a tax preference add-back for purposes of computing the corporate alternative minimum tax.
- Losses on contracts can be recognized for tax purposes when the contract is complete. Meanwhile, the contractor's financial statement must recognize the entire loss at the time that the contractor knows the contract will lose money.

Determining the Contract Completion Date: The law generally follows the premise that a contract is not complete until the specified work is completed and accepted by the customer. The regulations provide guidelines for determining when a contract is complete. The major consideration in these regulations is the substantial completion requirement for recognizing income. Previously, there had been cases which held that completion meant final acceptance and completion of 100% of the contract. Substantial completion under the existing regulations could now be interpreted in many circumstances to occur from 95% to 99% completion, depending on the individual facts and circumstances.

The general rule is that a contract is complete for tax purposes when final completion and acceptance have occurred. The regulations specifically state "a taxpayer may not delay the completion of a contract for the principal purpose of deferring federal income tax."

The contractor must look at the facts and circumstances of each contract in order to determine if completion has occurred. The principal factor to consider in determining contract completion is the status of the relationship between the parties to the contract regarding the requirements of the contract. One indication is *whether or not final payment has been made*. If the owner has made the final payment, it must be assumed that he is satisfied that the contractor has completed the project. Another measure is the *contract completion date* specified in the contract. Any *use of the property* constructed under the contract (occupancy of a building, opening of a road, etc.) by the owner also generally indicates completion.

The lack of an occupancy permit or the absence of final release of retention does not necessarily keep a contract open, and final testing or completion of punchlist items are not clear factors allowing a contract to remain open. Contingent compensation does not delay final completion. Completion is determined without regard to any disputes that may exist at the time of final acceptance (i.e., claims do not keep contracts open).

Severing and Aggregating Contracts

The Tax Reform Act of 1986 includes a provision that gives the Treasury the authority to issue new regulations treating two or more contracts as one, and one contract as two or more separate contracts. In addition, current regulations have given the IRS broad powers to determine when contracts should be treated separately or aggregated. The determination is made based on the facts and circumstances of each case, with the ultimate decision based on the principles of substance over form.

The following list includes factors used to determine whether a contract should be severed, and the types of specific situations that result in two separate contracts.

- Does the contract contemplate separate delivery, or acceptance of units representing a portion of the contract?
- Are the units in the contract independently priced?
- Are there business reasons justifying one agreement rather than multiple contracts?
- What are customary commercial practices for the type of contract?
- What dealings have occurred between the parties to the contract?
- How much time is contemplated between delivery of each unit?
- What is the nature of the contract clauses?
- How many units are contracted for under the contract? Options to increase the units under the contract will be considered separate contracts.
- Are change orders often separate contracts (especially when they involve increasing the number of units beyond those specified in the contract).
- The postponement of a portion of a contract until a later date severs that portion of the contract. An example is the completion of the base building portion of an office building and the postponement of the tenant improvements until there are tenants. This situation would sever the contract.

Contract Claims

The law imposes a different set of rules on contractors for income recognized from a contract that involves a dispute. The rules differ depending on whether the contract is a *loss* or *profit* contract.

When a contract has been completed, but a dispute exists regarding either contract price *reduction* or *additional work* to be performed, the rules provide that all items of income and expense that relate to the disputed items are recognized only in the year the dispute is resolved. As a result, the contractor must defer claims costs until the dispute is resolved. However, there are two exceptions to these rules.

- If the dispute is so large and concerns such a great portion of the contract that it would be difficult to determine the profit or loss on the contract, then the *entire* contract should be deferred until the dispute is resolved.
- If the contractor is *assured of a profit* (cost plus fee) regardless of the outcome of the claim, then the contract must be closed in the aggregate in the year completed, regardless of any other factors. (The same rules apply for loss contracts.)

In cases where the dispute involves claims for an increase in contract, all costs and income are recognized in the year the contract is completed. The disputed amounts of income and any related costs are included in the year they are earned or incurred. Signed change orders should always be included in the contract price when closed. A dispute over payment of the contract price does *not* mean that the amount in dispute is not recognized.

The Corporate Alternative Minimum Tax Amount

Prior to the Tax Reform Act of 1986, there was an "add-on" corporate minimum tax which was limited to most regular corporations. Under the 1986 Act, the minimum tax computation was modified to act as a parallel tax system. This was achieved by making adjustments to taxable income and computing a separate tax liability. The resulting minimum tax liability is compared to the regular tax liability, and the greater of the two tax amounts is paid. This discussion includes only the alternative minimum tax for regular corporations, as the rules vary between individuals and corporations. However, all of the adjustments and preferences included for corporations generally also apply to individuals.

Individuals have available adjustments for passive losses and investment interest expense, excess itemized deductions, incentive stock options, research and experimentation expenses, and circulation expenses as additional tax preference items. These are not subject to the book tax difference add-back.

An example of the calculations used to determine the corporate alternative minimum tax (AMT) is shown in Figure 8.5. The calculation of corporate alternative minimum tax always starts with the regular, corporate taxable income. Depending on the individual item, there may be an addition, a substitution, or a recomputation of regular taxable income. The major adjustment items for arriving at alternative minimum taxable income are discussed in the following paragraphs.

Depreciation

For property placed in service prior to 1987, the difference between the *accelerated* depreciation and *straight line* depreciation is a tax preference addition to taxable income. Accelerated Depreciation includes any declining balance or Accelerated Cost Recovery System (ACRS) method. For property placed in service after 1986, an adjustment to the depreciation deduction is made. For real property, the life over which the depreciation is calculated is extended to forty years. For personal property, the depreciation is calculated using the 150% declining method over the asset's ADR class life. The ADR class life is generally longer than the regular tax depreciation life.

Income recognition

The percentage of completion method for calculating income on long term contracts entered into after March 1, 1986 must be substituted for the taxpayer's regular method, such as completed contract or percentage of completion capitalized cost method. Installment sales after March 1, 1986 are not allowed for AMT purposes; therefore, the entire gain is recognized in the year of sale.

Tax Exempt Interest

Tax exempt interest on private activity bonds issued after August 7, 1986 is taxable for AMT purposes. Generally, private activity bonds are those not issued by a government agency or body for any public purpose, such as roads or schools.

Net Operating Loss Deductions

Net operating loss carries forward subsequent to 1986 and must be recomputed on the same basis as the AMT.

Book Income Adjustment

A major item of tax preference is the new "book income" adjustment. The corporation's alternative minimum tax (AMT) is increased for one half of the excess of pre-tax book income over AMT. Book income is generally referred to as the taxpayer's *applicable financial statement*. The IRS has provided the following order of priority regarding the financial statements issued by a company:

- SEC statements
- Audited statements used for credit or shareholder purposes
- Regulatory financial statements
- Any other financial statement used for credit purposes or distributed to shareholders, including internally prepared statements for management's use.

Calculation of Corporate Alternative Minimum Tax	
Regular taxable income	$14,000.00
Add: Alternate minimum tax preference items, such as accelerated depreciation of real property, book income adjustment	
Adjust the regular taxable income for items requiring a differing accounting method (i.e., substitute percentage of completion accounting for completed contract accounting)	45,636.00
Less: Alternative minimum tax net operating loss carry-overs	( )
Less: Allowable exemption amount	(40,000.00)
Alternative minimum taxable income	19,636.00
× 20% alternative minimum tax rate	3,927.00
Less allowable investment tax on foreign tax credits	( )
Subtotal	3,927.00
Less regular tax liability base on regular taxable income	(2,100.00)
Alternative minimum tax liability	1,827.00

Figure 8.5

The adjustments made to the amount of book income include nonconsolidated entities for tax purposes, federal, or foreign taxes. Extraordinary items reflecting net of tax must have the tax effect removed.

Corporations are allowed a $40,000 income exemption for AMT purposes. However, this exemption is phased out at the rate of 25 cents for every dollar of AMT over $150,000. Investment tax credits that carry forward from prior years may offset 25% of alternative minimum tax, but no other tax credits are allowed against the AMT.

If a corporation is subject to the alternative minimum tax, then the *excess* of the AMT over the corporation's regular tax is a *credit*, carried forward and used to offset regular tax in excess of alternative minimum tax in future years. This tax credit is designed to eliminate double taxation of the same income.

Planning for the Corporate AMT

Tax planning for construction contractors is significantly affected by the alternative minimum tax. Because it is virtually impossible to completely eliminate the corporate tax liability, it becomes more important for taxpayers to *plan* tax payments, to minimize their impact on the firm's cash flow. The basic goal should be to insure that the corporation has regular tax liability in each year. This liability should slightly exceed the alternative minimum tax liability, thereby minimizing its impact. This calculation is commonly referred to as the AMT *breakeven*, or *crossover point*.

Consideration should be given to the purchase of heavy equipment, as the AMT adjustment for depreciation may cause unanticipated taxes resulting from slower depreciation "lives". Many contractors may want to consider changing their income recognition accounting methods for either tax reporting or financial reporting purposes. In this way, they may reduce the complexity of tax calculations and applications. "S" corporations offer a further advantage in that they are not subject to the book income adjustment.

Asset Depreciation

In the past, contractors sometimes ignored depreciation as a business expense on heavy equipment. They believed that well maintained equipment created additional wealth for their company during periods of moderate or high inflation. With the moderation of inflation in the 1980's, this theory may no longer be valid.

In 1981, Congress modified the tax depreciation rules for equipment and real estate to create a capital investment incentive through the use of accelerated tax write-offs for fixed assets. This act created an entirely new term for depreciation: ACRS, or *Accelerated Cost Recovery System*. The ACRS had significant importance to the construction contractor in the following way. It allowed annual depreciation deductions for tax purposes in excess of the economic depreciation of the equipment. The results were a tax savings and a reduction in the investment cost for the equipment.

The Tax Reform Act of 1986 modified the ACRS depreciation system. The period of time for the write-offs and the methods available are summarized in Figure 8.6. Fixed assets are put into classes and depreciated according to their recovery life. Generally, contractors find that most equipment falls into either a three, five, or seven year class life.

An annual "bonus" depreciation deduction of $10,000 is available to the small contractor for assets bought during the year. The requirements for taking this additional deduction include the following. Income from the contractor's business must exceed the amount of the deduction, and the business must have purchased less than $200,000 of fixed assets during the year. Contractors anticipating large equipment purchases in excess of the $200,000 limit should try to "straddle" the purchases over two years in order to stay below the $200,000 level in any one year.

Modified Accelerated Cost Recovery System

Asset Type	Recovery Method (Depreciation Method)	Recovery Life (Depreciation Life)
ADR midpoint life of 4 years or less, excluding autos and light trucks. Includes over-the-road tractor units	Double declining balance rate of 66.67%	3 years
ADR midpoint life of more than 4, but less than 10 years, incl. light trucks and cars, heavy general purpose trucks, computers and peripherals, office equipment	Double declining balance rate of 40%	5 years
ADR midpoint life of 10 years, but less than 15 years, such as office furniture	Double declining balance rate of 28.7%	7 years
10-year	Double declining balance	ADR midpoint life of 16 years and less than 20
15-year	150 percent declining balance	ADR midpoint life of 20 years and less than 25 years, plus: sewage plants
20-year	150 percent declining balance	ADR midpoint life of 25 years or more, other than real property with ADR life of 27.5 or longer
27.5-year	Straight-line	Residential rental real estate, elevators and escalators
31.5-year	Straight-line	Commercial Real Estate

Figure 8.6

Real Estate Investments

No discussion of tax management would be complete without including real estate. It seems that no contractor is satisfied until he has been able to participate directly in some real estate investment and development activity. Until recently, congress has historically favored real estate investment with beneficial tax treatment, such as the 19-year ACRS (Accelerated Cost Recovery System) depreciation. However, as part of the Tax Reform Act of 1986, Congress substantially modified the tax treatment of real estate. This act has limited the benefits of developing and owning real estate. The specific changes were numerous and are too complex to discuss in the context of this book. In summary, they included:

- Real property depreciation is limited to either a 27.5 year life for residential, or a 31.5 year life for commercial real estate based on a straight line method of calculation. The result is a significant reduction in depreciation deductions for owners of real estate.
- The uniform capitalization rules for costs include all construction projects and eliminate almost all development period expenses as current deductions. Therefore, real estate investments generally do not begin to produce any tax losses until they are in use.
- Rental real estate activities have been characterized as *passive*. The tax losses from such activities are deductible only against income from other passive activities. These rules are subject to certain exceptions for taxpayers with less than $150,000 of income who "actively" participate in the rental activity. There are also some phase-in restrictions.

The result of these changes has been to limit investment in real estate to those developments that can produce economic returns commensurate with the risks involved. Investment is often limited to entities that can afford to invest at very low rates of current return in anticipation of higher long term returns.

Local, State, and Federal Taxes

State and Local Income Taxes

Most states levy a separate income or franchise tax. Many cities and other local jurisdictions have now implemented an income tax. Contractors doing business in multiple states, or a corporation with large income from investment or other corporate sources, should pay attention to the incorporation and allocation methods utilized by the various states.

Many states and localities also impose excise taxes; taxes on capital stock or business license taxes, and additional payroll-related taxes. The costs of these local taxes should be evaluated prior to bidding a job in an unfamiliar location.

Federal Withholding and Employment Taxes

Most contractors are familiar with the fact that federal withholding tax, FICA, and unemployment taxes are part of doing business whenever the contractor has employees. The one word of caution here is that contractors, like most small businesses, are tempted to defer the payment of these taxes to the government when cash is tight. There are substantial penalties and interest involved in making late payments and filing late payroll tax returns. If the corporation goes bankrupt, the officers of the corporation cannot escape liability for nonpayment of these taxes.

Federal Excise Taxes

For contractors involved in road work and transportation of materials, federal excise taxes on fuel and trucks are always a consideration. This expense should be included in the firm's financial planning.

Sales and Use Taxes

Many contractors experience an audit of their records by a state sales and use tax agent long before they are ever audited by the IRS. States and localities depend on these taxes for a large portion of their revenues. Because contractors deal in large dollar volumes in construction projects, they are important targets for state revenue agents.

Although the laws vary from state to state, generally when a contractor purchases materials, supplies, or equipment, a sales tax is levied. Most states impose *use taxes* on purchases made out-of-state if the state in which the material was purchased was not able to collect taxes.

Numerous cost-savings techniques can be used to minimize sales and use taxes. However, because of the diversity of state and local laws, coverage of those techniques is beyond the scope of this book. Some of the more common areas to be evaluated for possible sales tax savings include resale goods, tax on rentals, service work, labor and maintenance contracts, and the labor included in fabrication and installation contracts.

Fixed Assets

Taxation plays a major role in management decisions pertaining to the purchase or leasing of such assets as field equipment, office facilities, or office equipment. As discussed in Chapter 6 on Asset Management, all cash inflows or outflows generated by fixed assets during their useful life are subject to either additional tax costs or tax savings.

Other Tax Considerations

The following are additional tax issues that should be considered in the financial planning of contracting companies.

- **Employee vs. company-owned vehicles.** The use of employer-owned vehicles by employees can create both a benefit for the employees and a record keeping nightmare. As an alternative, the contractor may want to consider making agreements with employees for use of their own personal vehicles.
- **Retirement and Fringe Benefit Plans.** Fringe benefits are an important factor in retaining qualified employees. Although retirement plans have become and will continue to be more restrictive, they do provide an incentive for many employees. Contractors should also evaluate the use of other fringe benefits such as cafeteria plans, group life insurance, disability insurance, health insurance, and other benefits that provide nontaxable compensation to themselves and their employees.
- **Estate Planning and continuity of the business.** Contractors, like all other people, would prefer to think that they will never need to carry out estate planning. However, proper estate planning can make the difference between a company's continued existence after the owner passes away and a forced sale at a "fire sale" price. Planning for the smooth succession of management and ownership should begin early in the existence of any business.

Conclusion

The separate tax considerations that apply to the construction contractor are almost too many to count. Added to this complexity are the effects of the Tax Reform Act of 1986. This act not only reduced the availability of tax deferrals to all contractors, but also added substantial record keeping and tax planning burdens to the business. The new income reporting requirements, alternative minimum tax, and limitations on corporate fringe benefits have increased the probability that contractors will pay substantially higher taxes, beginning with 1987. For many contractors, the additional administrative cost and professional fees to manage their tax situations may be an incentive to change their tax accounting methods, reducing their complexity in future years.

Most tax planning is aimed at accomplishing one goal: deferring taxes. The cost of these deferrals must, however, be measured against their benefit. It becomes clear that the contractor cannot begin to manage this process effectively without the help of competent advisors. The selection of experienced advisors—CPA's and lawyers—should always be the contractor's first step in the tax management process.

Chapter Nine

AUTOMATED FINANCIAL APPLICATIONS

Chapter Nine

AUTOMATED FINANCIAL APPLICATIONS

During the last ten years, computers have become one of the most productive tools available to the construction industry. While computer hardware has become more affordable and better quality, a wide variety of industry-specific software has also been developed. Even small firms with less than $1 million in annual sales have found that computers offer improved efficiency and control well worth the investment.

The reason for the rapid growth in computerization is quite simple: computers can perform many routine tasks that previously required a great deal of time and effort from accountants, project managers, and bookkeepers. Efficient use of computers leads to improved internal control and profitability and provides managers with more time to focus on other business issues, such as planning, productivity, and decision making. While computer systems cannot provide the human reasoning and judgment so crucial to the management process, they can efficiently perform the following functions.

- Accept, store, and retrieve information.
- Process mathematical calculations.
- Perform complex analytical and statistical procedures.
- Display and report the status of information.

Computerization offers contractors many advantages in control, efficiency, and profitability. Several of the most direct and valuable benefits include:

- Formalization and "streamlining" of the accounting process.
- Easy preparation and access to comprehensive financial reports, such as the firm's balance sheet and income statement.
- Up-to-date job cost reports, formatted so that management can quickly recognize and react to problem areas.
- Accurate and rapid estimating of direct project costs.
- Greater control of accounts receivable, allowing rapid follow-up on collections and, consequently, improved cash flow.
- More efficient control of subcontractors and vendors in terms of insurance requirements, payments, change orders, and extras.
- Preparation and/or tracking of purchase orders, assuring accurate job cost control and minimizing the possibility of erroneous payments.
- Improved understanding of the costs and profitability of investments in equipment and inventory.

Few businesses generate and process more financial data than construction firms—data that must be collected, processed, maintained, and easily retrieved. Applying computers to the firm's accounting functions offers the most direct results in accuracy and efficiency. However, many other administrative and project management functions can also be computerized. (See Figure 9.1.)

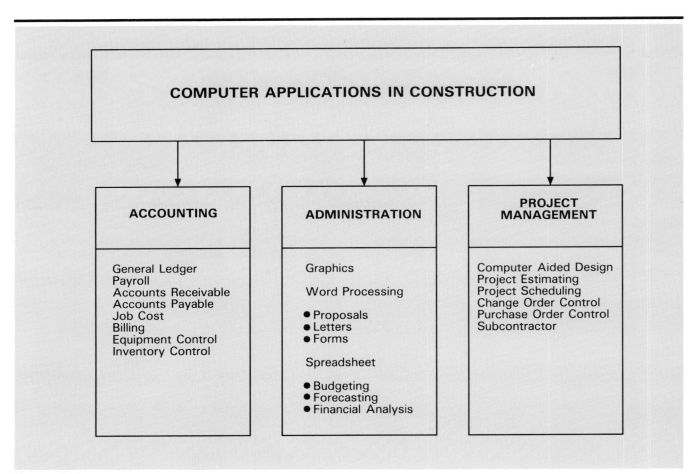

Figure 9.1

Management Information Systems (MIS)

Computers are the tools used in *Management Information Systems* (MIS). As shown in Figure 9.2, an MIS requires input from managers and records of business data. The system outputs are various financial and managerial accounting reports. The purpose of an MIS is to support managers by providing ready access to business data in a usable form to help solve business problems. The output from a good MIS is well organized, accurate, and rapidly accessible.

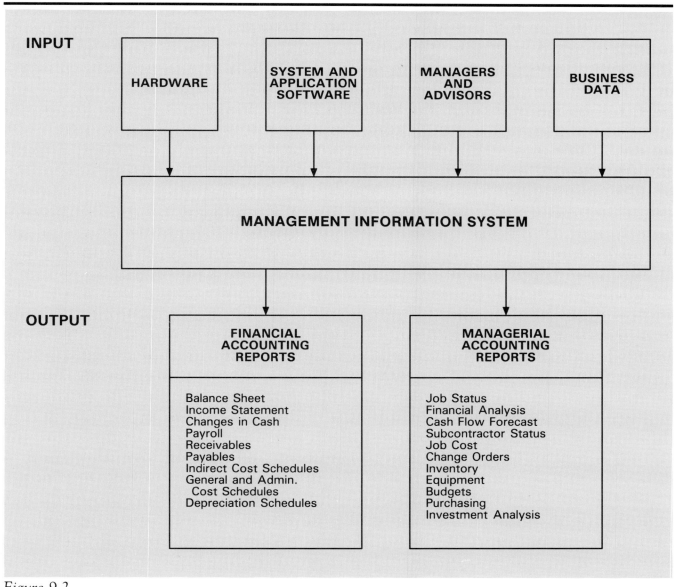

INPUT

| HARDWARE | SYSTEM AND APPLICATION SOFTWARE | MANAGERS AND ADVISORS | BUSINESS DATA |

MANAGEMENT INFORMATION SYSTEM

OUTPUT

FINANCIAL ACCOUNTING REPORTS

Balance Sheet
Income Statement
Changes in Cash
Payroll
Receivables
Payables
Indirect Cost Schedules
General and Admin.
 Cost Schedules
Depreciation Schedules

MANAGERIAL ACCOUNTING REPORTS

Job Status
Financial Analysis
Cash Flow Forecast
Subcontractor Status
Job Cost
Change Orders
Inventory
Equipment
Budgets
Purchasing
Investment Analysis

Figure 9.2

By integrating the various financial and project management applications into a Management Information System, it is possible to attain managerial control and efficiency not available through conventional manual systems. Given the complex and competitive nature of the building industry, such immediate access to accurate and current information may make the difference between profits and stability, versus losses and bankruptcy. It should be noted, however, that a construction firm must have an MIS fully established and operational before automating. For, as has often been said of computers, "garbage in, garbage out."

Computer Hardware

While an extensive discussion of construction accounting software is included in this chapter, it is beyond the scope of this book to explore computer hardware in great depth. This topic is so complex and ever-changing that numerous books and periodicals are published to keep pace with this dynamic field. A brief discussion follows, covering a few technical subjects and major factors to be considered in selecting computer hardware.

The term *hardware* refers to the computer equipment itself, including the video display, printer, and keyboard. An example of a specialized hardware device is a *modem* that allows computer information to be transferred over telephone lines. This may be an important feature if any accounting functions are performed in the field. The *keyboard* resembles a typewriter and is used by the operator to enter the data. The *video display terminal* is similar to a television screen and displays the entries. The user can view the data as it is entered, as well as the effects of processing that data. The *printer* produces a "hard" (paper) copy of the work the computer performs, such as the general ledger, job cost reports, financial statements, and checks.

Memory and Storage Capacity

The computer has a capability referred to as *memory*. Computer memory is a physical means for storing electronic pulses that represent certain information. There are two basic types of computer memory: ROM and RAM. ROM stands for *Read Only Memory*. It is an area of electronic circuitry that stores a set of instructions the computer can read and act upon, but cannot change. For example, ROM stores the instructions necessary for the computer to begin operation once the power is turned on, and other sets of instructions required for the computer system to operate.

RAM, or *Random Access Memory*, is circuitry that temporarily stores electronic pieces of information. The computer can *write* information to RAM, manipulate it, change it, or erase it. Because it is only a temporary storage area, all information stored in RAM would be lost once the computer is turned off. We can save any data placed in RAM by storing it as a file on some type of magnetic device. This could be a floppy diskette, a hard disk drive, or a magnetic tape.

The capacity of both memory areas and storage devices is measured in *bytes*. One byte represents one alphanumeric character. For example, the letter A is one byte in size. The number 523 is three bytes. Each space, period, numeral, or character occupies one byte. One thousand bytes are referred to as one *kilobyte*, or one *K*. One million bytes are referred to as a *megabyte*, or one *Mb*. Most floppy diskettes are 360 K in size, which means that each can store about 180 double-spaced typewritten pages of information. The more durable and expensive *hard disk* can store thousands of pages, with a capacity of hundreds of megabytes.

Most systems installed today for contractors are either micro- or minicomputers, as opposed to mainframe computers used for very large applications. Microcomputers are small single-user and multi-user machines, often called *personal computers*. While microcomputers have been used primarily for such tasks as word processing and spreadsheet preparation, software vendors are now offering construction accounting, scheduling, and estimating software that run on micros. Larger contractors using a wide variety of applications may need multiple terminals. They generally select the powerful mini systems, capable of operating many terminals simultaneously with different applications of the same data.

Operating Systems

When people refer to personal computers, they normally are talking about computers running in a single-user mode under the MS-DOS operating system. However, a group of independently operated PC's can be connected and, possibly, made multi-user. Special systems are available to allow this "networking". Networking enables users, with the appropriate application software, to use MS-DOS (a single user operating system) in a multi-user mode. This is not the *optimum* way to operate in a multi-user mode, but it does enable the user to accommodate MS-DOS programs to a multi-user environment. Other specialized software can be installed either on the IBM PC/AT or on an 80386 processor to obtain an efficient multi-user mode.

An 80386 machine offers the following additional advantage:

- it may be considered a personal computer when running MS-DOS
- it may be connected to a multi-user network of personal computers
- it can serve as a multi-user microcomputer when operating under specialized operating systems

The distinction is fading between personal computers and multi-user computers. In fact, some microcomputers are as or more powerful than minicomputers. The basic difference between a personal computer and a multi-user microcomputer is the way in which the computer handles the disk Input/Output (I/O). Multi-user microcomputers are more adaptable to a multi-user environment than personal computers because of their efficiency in handling disk Input and Output. Even though a personal computer with an 80386 processor has processor capabilities equivalent to that of a multi-user microcomputer, the disk handling features are normally so poor that they cannot handle as many terminals in a multi-user mode as an efficient multi-user microcomputer.

Selecting Hardware

A wide range of factors, other than price, should be considered in evaluating computer hardware. First and foremost, the buyer should *ensure that the desired software will operate on the machine*; the selection of software must always precede the choice of hardware. The second major issue is the *quality and reliability* of the hardware, as well as the *terms for service, technical support*, and *experience offered by the vendor*. The system should have the *capability to expand* as the firm grows. Expansion may involve additional memory and storage; additional software applications; new peripherals, such as modems and specialized printers; and additional terminals to allow multiple users. Ideally, all stored data and existing equipment should be retained as expansion occurs, minimizing the impact of future system upgrades.

The choice of an interactive or batch system is another factor to consider. An interactive system provides current data to all software modules instantly, as each piece of information is entered. The batch system, on the other hand, does not update information as it is entered. It instead processes each program in a batch after all data has been entered. Finally, consider the need for single-task processing versus multi-task processing. The latter allows the handling of many jobs and applications simultaneously, which is a requirement in larger firms with multiple users and tasks.

Computer Software

Computer software is a set of predefined specialized instructions that direct the computer to organize information, correlate data, and accomplish a specific task or tasks. Software instructions that operate at the machine level are referred to as *assembly language*, and vary in programming depending on the microprocessor. The *operating system* software coordinates the input, output, reading, writing, and printing functions. The operating system also includes *utility* programs for daily "housekeeping" functions such as the backup, copy, and movement of files.

Software programs are developed by professional programmers to perform specific functions, such as accounting, scheduling, or estimating. Different computers recognize different operating systems; therefore, software programs must be written to accommodate each type of computer system. It is always advisable to purchase the *most suitable software* program first, and then the hardware system.

In the past, many larger contractors have developed custom programs to meet the needs of their companies. This approach can, however, be costly and frustrating. Fortunately, standard, industry-specific programs are now readily available to meet the specialized demands of the construction industry. Many such core programs can be further developed and custom-enhanced to satisfy the special needs of any given contractor. The advantages of these programs include low cost, availability of full documentation, program maintenance by the vendor, and portability among different brands of hardware.

Two crucial factors in evaluating software are the levels of interfacing and integration possible. *Interface* refers to the sharing of a common data base by all modules, simplifying operation and minimizing storage space. *Integration* refers to the interaction between modules that allows a single point entry to automatically update all affected modules. For example, when a cost that is related to a job in progress is entered in Accounts Payable, the program should automatically post all affected modules, such as the General Ledger, Job Cost Ledger, and purchase orders.

Selecting Software

The features and associated costs of computer software vary dramatically, as do the needs of contractors involved in different types of projects and with varying sales volume. The task of matching up the ideal software with the needs of a specific firm may well demand the services of an independent consultant. The selection will have a tremendous impact on the long range efficiency and profitability of the firm, so time and cost should be secondary considerations. It is not unusual to spend several months gathering data from different vendors, analyzing the basic and special characteristics of each software package, negotiating terms, and making a decision.

It is advisable to start the process by developing a Management Information System (MIS) Mission Statement, outlining the overall goals to be achieved in implementing the computer system. Key factors to be analyzed include:

- What goals does the firm expect to accomplish through the MIS, and do the benefits justify the cost?
- What is the current annual sales volume and anticipated growth over the next three to five years, and how will the growth affect the needs of the firm as it relates to the MIS?
- What are the current methods of processing and reporting information, and how will they or should they change with the MIS?
- Who will operate and maintain the system, and how will these individuals be trained and supported?
- How will the new system be implemented, and what process will be used to develop the appropriate data base?
- In what order should the various parts of the business be computerized, and what time frame should be allowed for each phase?
- What kinds of hardware and software are needed to accomplish the goals and intentions specified in the Mission Statement, and how will they be selected?

Once equipped with a clear MIS Mission Statement, it is necessary to become familiar with computer systems in general regarding cost, features, reliability, and service. Ideally, a Request For Proposal (RFP) may be prepared at the early stages so vendors may respond to the specific requirements of the firm. Vendor references, history, commitment to the industry, and financial stability should be carefully checked. It is crucial that the vendor be available for total support on both the hardware and software, including implementation, training, updates, and maintenance. Other important factors to consider in evaluating software are:

- Initial cost, as well as cost of annual technical support and new updated versions provided by the vendor.
- Reliability of all functions as stated, without errors or bugs that prevent smooth operation.
- Usability, or ease of use, in terms of data entry methods, consistent use of terms and procedures, understandable diagnostic messages in the case of operator errors, speed of operation, and clarity of manuals.
- "Crashproof" properties, or the resistance to detrimental results from striking the wrong key.
- Language used and the associated portability among computers.
- Use of menus and operator prompts to provide user clarity and to simplify operation.
- Availability and flexibility of password protection. Ideally, password protection can be placed on any given program or file, and can be readily changed.
- Convenience and reliability of back-up procedures, that provide duplicate copies of data and programs in the event of loss of electricity or equipment failure, operator error, fire, vandalism, or other problems.
- Limits of data lost in the event of a power outage. Ideally, only those transactions that appear on the screen at the time of the outage are lost.

- Access to the software source code that enables the buyer to perform maintenance or program changes if the vendor terminates support.
- Accessibility to central files by all users, so all the data can be maintained through a common data base structure.
- Ability to "spool" reports, or select and print multiple reports without further intervention from the operator.
- Quality of the vendor proposal, including the specific terms, services, and responsibilities of both parties.

This is an extensive list of issues to consider when selecting software. It may be helpful to draw up a comparative chart, that will reveal the significant differences between packages. At this stage, two or three packages are apt to emerge as favorable candidates and demonstrations should be requested. Such presentations provide an excellent opportunity to experience the actual operation of the system and to ensure that it is as functional and effective as the literature and vendor have indicated.

After selecting a software system (but before informing the vendor that a decision has been made), a contract is typically prepared covering the terms of the purchase, installation, system performance, and follow-up service. Many vendors have standard contracts, but these are often one-sided and should be reviewed by an attorney. The following specific features should be included in the contract:

- A listing of all equipment and software to be provided, as well as all associated warranties.
- The level of assistance with installation and training to be provided by the vendor.
- Delivery dates, prices, and payment terms.
- Cost and terms of maintenance and follow-up support.
- Time frame and requirements for implementation and testing, allowing one final acceptance test.
- List of reports to be generated and processing time required.
- Availability and cost of future versions of the software.
- Penalties or remedies for nonperformance, including arbitration.
- Rights of cancellation, termination, or resale.

System Installation

The key to the smooth installation of a Management Information System is preplanning, using the MIS Mission Statement. The management staff in general, and operators in particular, need to be fully informed of internal changes that are to take place in the flow of paperwork and individual responsibilities. Primary as well as back-up operators must be well trained in advance, hopefully by the vendor providing the system. During this preparation and training period, it may be desirable to enlist temporary help to handle routine office work, allowing the regular office staff to concentrate on getting the new system up and running as smoothly as possible.

The location of the equipment must be considered in terms of six factors: traffic flow, utility connections, environmental conditions, storage space, security, and the work station itself. Because the financial function is typically the first to be computerized and is apt to be used most heavily, most systems are generally incorporated into the accounting department offices. However, due to the wide variety of uses and hardware configurations, these six locational factors must be individually analyzed for each firm.

Using the MIS as a guideline, a manual outlining the company's managerial and financial control policies and procedures can be developed, defining the firm's internal structure and operational methods. This manual should be brief, simply written, and clarified with flow charts and other diagrams. Procedures and parties responsible for purchase orders, invoices, time cards, and other source documents should be included. Procedures should also be outlined for processing the general ledger, job cost, payroll, and financial reports.

Knowing the current information handling system, management can determine what it expects to achieve with the new computer system. This analysis should include a plan for simplifying the specific functions most in need of more accuracy and speed, as well as the format and use of various reports. Specific responsibilities, both during and after the start-up period, must also be clearly assigned. Once this analysis is complete, the procedure manual can be completed, outlining all instructions for operating the new system and handling documents. The manual should be carefully followed by all personnel, with refinements regularly made and communicated.

A base date, or the starting point for implementing the new system, should be established when all accounting data can be current and properly formatted for entering into the system. This information will include general ledger data, lists of current subcontractors and suppliers, outstanding invoices with general ledger and job cost codes, current clients, payables and receivables, and active jobs with associated budgets. Once this information has been entered and the system is up and running, it may be advisable to continue operating the existing system, in parallel, until it is clear that the new system is functioning properly—perhaps for several months!

Financial Control Software

Many construction accounting software systems are available from vendors. Some of these systems offer special features and are designed for certain types of contractors. The financial applications most often computerized by construction firms of all sizes include payroll, general ledger, job cost, and accounts payable. As a firm grows, it can easily justify adding computer applications for accounts receivable, billing, and purchasing. As more field equipment and inventory are acquired, procuring specialized modules to monitor the costs in these categories may be justified. Simple electronic spreadsheet programs may also be added early on for budgeting and financial analysis, allowing any number of "what if" calculations to be computed almost instantly.

In addition to the software selection factors previously discussed, other significant features generally desirable in accounting software are as follows.

- The program should automatically edit all data that is input in order to verify codes, projects, and other information. Only equal transactions, i.e., debits and credits that balance, should be accepted. The system should also allow for the transfer of data among files and modules.
- All entries should be available for screen inquiry or printed reports prior to posting. Each transaction should be tracked by audit trails, ideally indicating its history and the operators involved.

- The system should allow user-defined reports in detailed or summary form, using data from any file, and should recall data from a specified time frame. This selective reporting capability is particularly important for the specialized managerial accounting needs of different contractors. Account sorting at various levels is also important.
- The number of accounts, departments, jobs, subcontractors, and employees should, ideally, be limited only by storage space and not by the software.
- The system should allow account codes to be either alphabetical or numerical. Historical data should be retainable even though an account may be declared inactive and not shown on current reports.

As previously stated, all accounting modules should ideally be fully integrated to eliminate the need for double entry and updating. Figure 9.3 charts the relationship among the various modules in a completely integrated and full function contractor's system. Not shown are a few very specialized modules not normally required by most contractors, covering such areas as *warranty* and *fixed asset control*. The power and efficiency of the system shown in Figure 9.3 can be truly astounding if it is equipped with the features described in the following sections. There is, however, a direct relationship between the system's features and its cost. Therefore, each firm must perform a careful cost benefit analysis as part of the software selection process.

General Ledger

The general ledger lies at the heart of all automated accounting systems. It enables the firm to maintain a chart of accounts, make bookkeeping entries, and print financial statements. Virtually all activity data entered into any system module should automatically update the general ledger module and appear, in summarized form, in the general ledger module. Controls should be built into the general ledger module to insure the accuracy of the transferred data and agreement with all other modules. Many systems are capable of making such summaries not only on a company-wide basis, but also at the divisional or departmental level. Some packages are also capable of operating multiple companies at one time, a feature that may be important to growing firms managing property, developing real estate, or overseeing specialty trades.

Integration
The general ledger should be automatically updated with information provided from the accounts payable, equipment, inventory, job cost, payroll, and accounts receivable modules. In some systems, data from the general ledger can be directly input to the word processing files for ease in preparing various written reports that include the firm's summarized financial information.

Formatting
Most systems allow a user to custom format a chart of accounts to directly suit specific needs. This capability allows different cost centers or departments to be grouped in various ways, with budgets created for

each. Financial reports such as income statements and balance sheets should also allow flexible formatting, so that actual figures can be compared to figures from other reporting periods, or to budgeted figures.

Accounting Methods
In addition to cash base accounting, the system should also accommodate *accrual basis accounting*, which is necessary for good management reporting. End-of-period contract adjustments should be possible based on percentage of completion for each project or completed contract, at the user's option.

Open Period Processing
Many systems allow posting of a new month's or new year's activity prior to closing the previous period's activities. As a result, 24 months or more may be open at any given time. Without this capability, extensive delays may occur in the bookkeeping function while waiting to receive and process all previous period invoices and close the period.

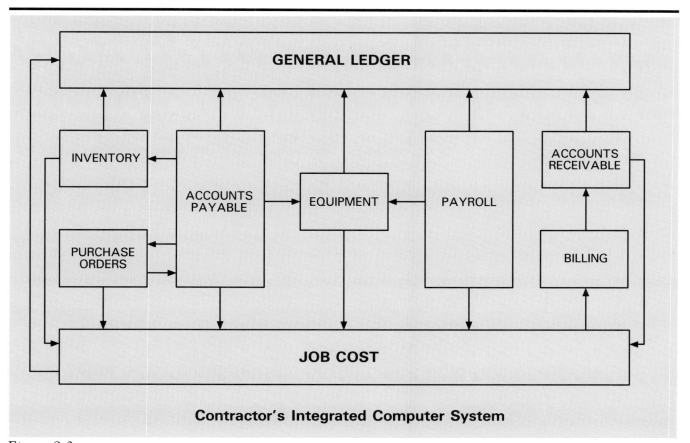

Figure 9.3

End-of-Period Closeout

This feature allows automatic monthly closeout and year-to-date balances to be brought forward for the next month's activity, once all posting and financial reporting is completed for the month. At the end of the fiscal year, closing entries to the profit and loss accounts should be made automatically to "zero" the balances of each account in preparation for the following year. Many systems allow for automatic recurring journal entries within each period and end-of-period audit adjustments. This feature can be an invaluable time saver when it comes to such routine entries as depreciation.

Reports

Major reports generated by the general ledger module typically include the following:

1. **Trial Balance and Trial Balance Worksheet**—displaying each general ledger account, balances brought forward for each account, all current period posting, subtotals for the current period, and the new balance for each account.
2. **Journals**—including the Sales Journal, Purchase Journal, Cash Receipts Journal, Cash Disbursements Journal, Payroll Journal, and General Journal.
3. **Chart of Accounts**—listing all accounts with numbers or codes.
4. **Financial Statements**—Income Statement, Balance Sheet, and Statement of Sources and Uses of Funds. Ideally, each of these reports should be available. Updated financial reports should be available at any time, factoring in percent complete adjustments and entries to date for current receivables, payables, and committed costs.
5. **Subsidiary Ledgers**—unlimited number of user-defined supporting schedules allowing analysis of major areas of concern through the isolation and sorting of specified income and balance sheet accounts, e.g., equipment, loans, tenant improvements, depreciation, cash in bank, cost of sales, balances forward.
6. **Financial Analysis**—an excellent feature offered by some systems, financial analysis provides ratios of key balance sheet and income statement data. As an alternative, such data may be loaded into a user-defined spreadsheet program which performs the calculations and prints the results.
7. **Transaction Register**—listing the transactions logged during any given period. This report allows easy examination of any transaction to ensure accuracy.

Accounts Payable

The accounts payable tasks include issuing checks and monitoring invoices and cash requirements. Most of the general ledger and job cost activity is generated from the *accounts payable* module, where source documents obligating the firm to future payments are initially entered. Therefore, the reliability, efficiency, and usability of the *accounts payable* application are critical. Many systems allow the user to make disbursements and to monitor both cash flow on a job-by-job basis, and subcontract activity. Some systems also allow the tracking of use tax for exempt and nonexempt jobs, as well as for selective materials within a particular job.

Integration

The *accounts payable* module automatically updates the general ledger, job cost, and equipment modules. It also updates and receives information from the *purchase orders* module.

Invoice Posting

If an invoice received is for materials used on several different projects, the system should post distributions to multiple jobs and accounts within the job cost module. Costs should be posted to job cost ledgers when the invoices are received rather than when checks are printed, regardless of the accounting method used.

Entry Verification

Most systems verify general ledger account numbers, job numbers, and vendor codes at the time of entry. Errors should be flagged so corrections can be made immediately. Inconsistencies between invoices and purchase orders should be flagged during entry.

Insurance

The *accounts payable* module will ideally monitor the expiration date of worker's compensation and liability insurance for all subcontractors, and will not print checks unless both are current. The system may also have the capability to deduct worker's compensation coverage from subcontractors who are lacking it, if so desired by the operator.

Checks

The accounts payable module should print vendor checks in bulk or individually. It should accept information from handwritten checks, while drawing on any number of banks and checking accounts. The operator should have the option of selecting, holding, or making partial payments on any invoice by vendor or transaction. Payments should be made on the basis of invoices or purchase orders, always checking for duplication or numerical errors. The operator should be able to view disbursements on the screen or imprint prior to the printing of checks.

The system should preclude payments to subcontractors or vendors in excess of approved contract amounts or purchase orders. It should also log and monitor change order payments. In addition, the system should handle retainage, backcharges, discounts, prepaid invoices, the voiding of single or multiple checks, and invoices that recur on a regular basis (such as rent and note payments). Another special feature available in some systems is the ability to select and pay grouped invoices according to those with discounts, for a particular project, with a particular due date, or for invoices individually selected.

Vendors

Ideally, the system will handle any number of vendors, limited only by storage space. Information in the vendor master file typically includes the vendor name, address, phone number, current balance, last payment date and amount, aged balances, year-to-date invoices and payments, discount information, back charges, comment from the field, and retainage information.

Reports

1. **Purchase Order Ledger**—itemizing outstanding purchase orders by number for each vendor, amount quoted, job and category of cost, and any payments to date.
2. **Open Payables Report**—itemizing unpaid invoices for each vendor with invoice number, purchase order number, and distributions made to the general ledger and job cost module.
3. **Government *1099* Forms**—outlining sums paid to each subcontractor on a yearly basis.
4. **Audit Trail**—chronologically listing all invoice entries, adjustments, sources, operator, codes, and other identifying information.
5. **Cash Disbursements Journal**—listing all checks written and corresponding entries in the general ledger accounts.
6. **Cash Requirements**—chronologically listing sums committed through invoices and purchase orders.
7. **Pre-check Report**—listing all invoices prepared for check issuance.
8. **Vendor Analysis**—listing the total contract amount, amounts billed, amounts retained, change orders, insurance status, and amounts paid to all suppliers and subcontractors.
9. **Use Tax Report**—listing all use taxes by job and by job cost account.

Job Cost

The purpose of the job cost module is to track expenditures and productivity for each building project, thereby allowing management to monitor and respond to current costs in comparison to the original budget. The system should also allow adjustments for change orders and cost updates without altering the original budget figures. Through effective use of the reports generated by this module, management can identify potential cost overruns early on and take the required corrective action. In addition, the reports provide one additional point of checks and balances on unexplained overpayments to vendors.

Integration

In fully integrated and full function systems, the job cost module will accept data from the accounts payable, equipment, inventory, accounts receivable, purchase order, and payroll modules. It provides input for the general ledger as well as the billing module if desired. Some systems are also integrated with an estimating application, allowing the budget for any given job to be directly loaded from the estimate module into the job cost module. In addition, historical cost data from previous projects can be used to load the estimate program for preliminary costing purposes on new projects. It should not be possible to log a cost in the job cost ledger without a corresponding entry in a specific project job cost file.

Formatting

User flexibility in formatting job cost codes for individual projects is important for contractors doing different types of construction. For example, a contractor involved in both commercial and residential projects is likely to accumulate and format job cost information in different ways for various projects. Subcontractors will also have far different job cost codes. Ideally, the system would be capable of accepting different classification formats for different jobs. The user should be able to define the level of report detail, for example, summarized costs by individual division, or detailed costs on each cost component within a division. A desirable feature is the ability to preclude the printing of cost categories with a zero budget, meaning that they do not apply to the project.

Cost Categories

Systems are available that are capable of classifying costs for individual line items into a number of categories, such as *labor*, *materials*, *equipment*, *subcontractors*, and *miscellaneous*. Within each of these classifications, the more sophisticated systems can track costs in terms of *dollars*, *labor hours*, *material units*, or *installed units*.

Change Orders

Some programs offer a change order control feature within the Job Cost module, allowing the contractor to maintain control over individual change orders applied to any given job. Change orders are retained in a separate file for review and editing prior to input into the job cost system. Change orders affecting subcontractors will be reflected in the Subcontractor Analysis Report of the accounts payable module.

Monthly Closeout

At month's end, the job cost module of many systems computes the percentage complete for each project by simply comparing the actual costs to date to the total current estimated direct costs for the project. The percent complete for each project is then compared to the percent of the total contract that has been billed to date in order to determine if the project is over- or under-billing. This data is then consolidated into one summarized general ledger adjustment to contract revenues earned for the period. In addition, all direct project costs should be automatically transferred at month's end from *Work-In-Progress* on the Balance Sheet to the *Cost of Revenue* on the Income Statement, assuming the firm is reporting on the accrual basis.

Reports

1. **Job Cost Report**—listing cost data for each line item as to original budget, adjustments, revised budget, committed costs (purchase orders), actual costs, and variances. This report should be available with full detail on all costs for each line item, or in summarized form.
2. **Contract Status Report**—outlining key financial data for each contract, such as contract amount, anticipated costs, earnings to date, earnings at completion, profits, retainage, and project receivables.
3. **Percentage of Completion Schedule**—listing all data pertaining to the over- and under-billing adjustments for each project.
4. **Change Order Log**—listing all change orders to date for each project with dates, amounts, and billing status.
5. **Exception Reports**—listing purchase orders or invoices that create over-budget conditions on any given classification.
6. **Productivity Reports**—contrasting actual unit costs versus estimated unit costs.

Payroll

The payroll module is designed to simplify and expedite the time-consuming functions of payroll preparation and reporting. Because of the wide variety of payroll variables involved in calculating paychecks for a firm's personnel, this is often the largest and most complex module within the system.

Integration

The payroll module updates the general ledger and job cost modules automatically. In the more sophisticated systems, the payroll module is also capable of updating an equipment module with mechanics' time, an important feature for those contractors who maintain their own heavy equipment.

General Features

Most payroll systems are provided with the following features:

- *Computation of total pay* at different rates for various individuals, and at different pay intervals. Pay can be based on an hourly wage, a pre-established salary, a commission and, in some cases, piecework. The system produces multiple computer-generated or manual checks per employee each pay period.
- *Distribution of all labor costs* and employer-paid insurances, payroll taxes, and fringe benefits to the general ledger and job cost modules. Appropriate entries should be made to accrued liability accounts for all accrued payroll expenses.
- *Miscellaneous one-time or multiple recurring deductions*, input individually for each employee. This feature allows for payroll advances, loans, and insurance contributions to be deducted on a selective basis.
- *Computation of Worker's Compensation* on the basis of the individual employee or the job function of the employee.
- *Entry of time sheet data* on a daily or weekly basis, at the discretion of the user.
- Multiple distributions of any single employee's *labor cost by job, phase, cost category, trade*, EEO category, Federal labor classification, and pay rate.
- *Audit trail and editing of records in payroll journal* for verification before posting to permanent records.
- Weekly, bi-weekly, semi-monthly, and monthly *payroll checks*.

Special Features

The more sophisticated payroll systems offer the following additional features:

- The ability to handle payroll on a multi-company, multi-division, and multi-department basis. This feature would be a requirement for any firm maintaining separate financial and managerial reporting systems for these various levels.
- Calculation of payroll taxes based on multiple state and local rates, by individual employee.
- Processing of multiple union deductions.
- Automatic distribution of indirect costs, including labor, to the job cost module.
- Multiple pay and charge rates for each employee within any payroll period, based on the employee, union, job, or labor category.

Reports

1. **Computer-generated Checks**—with all pay and deduction information listed on the check stub for employee and management records.
2. **Personnel Log**—listing pay rate, exemptions history, attendance record, bonuses, Worker's Compensation information, benefits, and special deductions for each employee. This report should also list the employee's address, social security number, marital status, date of birth, and tax codes.
3. **Withholding Reports**—listing withholding obligations of both the employer and employee for all taxes, health insurance, Worker's Compensation, benefits, unions, and special deductions.
4. **Government Reports**—printed in final form and ready for submission covering taxes, FICA (social security), FUTA (federal unemployment) and SUTA (state unemployment). Other special reports may be required by those firms engaged in government contracts. Among them are Equal Employment Opportunity, Certified Payroll, and Prevailing Wage reports.
5. **W-2 and 941 forms and reports.**
6. **Payroll Summary**—listing condensed or detailed payroll data by project.
7. **Union Reports.**

Accounts Receivable

The accounts receivable module facilitates the billing operation, monitors the firm's receivables and, ideally, prepares statements. It is integrated with the job cost module in such a way that over- and under-billings can be detected, and the general ledger can be consulted for updated receivable and cash receipts information.

Many systems manage retainage receivables separately from current receivables, an important feature for contractors. It may also be desirable to have miscellaneous billings and cash receipts separated from contract billings and receipts. Finally, some systems can track change order receivables from original contracts; this may be an important financial control feature.

The accounts receivable module is supported by a customer master file, that provides a data base of customer information to all reports. Data in this file should include:

- Customer name
- Contact person
- Telephone number
- Address
- Payment terms
- Aged receivables
- Discount percentage
- Markup percentage
- Total sales to date
- Retainage withheld from draws
- Billing history
- Service charges for late payment

Most accounts receivable modules are capable of processing and printing statements, although not always in A.I.A. draw format. Such statements will typically reflect the amount due, retainage, service charges, due date, and possibly special user-defined messages. Some systems are also capable of automatically issuing and recording recurring billings, a valuable feature for contractors or subcontractors with annual service contracts and the associated routine billings.

Reports

1. **Aged Receivables**—listing the customer, job, amounts billed, due dates, and overdue amounts.
2. **Accounts Receivable Ledger**—available with full detail or in summarized form.
3. **Billings and Cash Receipts Journals**—logging all current and historical activities.
4. **Sales Tax Reports.**
5. **Sales Reports**—listing current and historical billing information by customer.

Billing

While most accounts receivable modules offer a somewhat limited statement-processing function, the billing module available in many systems is more flexible and well adapted to construction firms. This billing module is integrated with the job cost module, receiving information to be used in processing bills, then feeding the data to the accounts receivable module.

Systems are available that produce invoices on the basis of *lump sum*, *time and materials*, and *unit price*, thereby addressing the needs of contractors operating on the basis of a variety of contractual terms. Options on billing procedure, calculations, and statement format must be user-definable for each job, allowing the terms of each contract to be addressed. Like the accounts receivable module, the billing module should also handle fixed or selective service charges, recurring bills, sales tax, change orders, retainage, and discounts.

In the case of time and material contracts, it is important to seek out a system that allows great flexibility in labor billing, including such methods as specified billing rates by trade classification, individual employees, and multiples of direct labor cost. This feature makes the task of labor billing automatic, not only saving time, but significantly reducing the potential for error. One of the most important features of the billing module for many contractors is the processing and printing of A.I.A. billing forms G702 and G703. (Copies of these forms appear in Appendix C.)

Once the user has input the schedule of values for each phase and the monthly percentages complete for each phase, the system will compute all current billing information as to total contract sum, change orders, portion of contract value completed to date, retainage to date, total previous payments, current payment due, and the balance remaining on the contract. Retainage can ideally be computed differently for completed work versus stored materials. The current billing and retainage data from these forms then updates the accounts receivable module accordingly.

Reports

1. **Billing Worksheet**—designed to simplify the task of preparing the percentage complete calculations for the A.I.A. forms.
2. **A.I.A. forms G702 and G703.**
3. **Invoices**—in a variety of user-defined formats.
4. **Billing History**—by customer.

Purchase Orders

The purchase order module is integrated with the job cost and accounts payable modules, and, in some cases, with the inventory module. The purpose of this module is to allow the user to enter and print purchase orders, record committed costs, process the receipt of materials, and assist in validating accounts payable invoices prior to payment.

In most systems, the user can be alerted when an invoice does not match a purchase order, when a duplicate invoice is received, or when an invoice is not supported by a purchase order. Ideally, purchase orders may be prepared for materials that are to be physically and financially distributed to a number of jobs, as well as to inventory. Another important feature is the ability to handle partial receipt of materials on any given purchase order and partial payments against total purchase order amounts.

In updating the job cost module, all job related purchase order amounts are listed in the job cost report as *committed costs*, and distributed in accordance with the associated cost code. The job cost report can, therefore, reflect not only actual expenditures to date, but costs not yet evidenced by an invoice or payment. This feature provides management with the most current job cost data allowing for corrective action at the earliest date possible.

Reports

1. **Purchase Orders.**
2. **Committed Cost Report**—listing all outstanding purchase orders by job with dates, amounts, and vendors.
3. **Invoice versus Purchase Order Report**—comparing the purchase order amount to the amount of the invoice(s) received for deliveries against the purchase order.

Equipment

The equipment module provides specialty functions useful to those contractors who own or lease their own heavy equipment. It enables the user to maintain cost and revenue data for each individual piece of equipment on a job-by-job basis or during any given period. It receives cost data from the payroll and accounts payable modules, and provides cost data to the general ledger and job cost modules. The equipment module should be able to handle multiple billing rates, depending on the type of equipment, hours used, or job. The costs are allocated to each job at time intervals established by the user, normally monthly. Costs of all types should be monitored, including loan payments, depreciation, operator wages, fuel, and maintenance.

Reports

1. **Equipment Log**—listing type of equipment, serial number, purchase data, price, depreciation to date, and present book value.
2. **Maintenance Report**—listing repair labor, parts, and schedule of required maintenance.
3. **Job Allocation**—identifying the jobs and the time the equipment has been used.
4. **Profit/Loss Report**—listing the revenues and expenses associated with each piece of equipment, as well as the level of utilization—for the current period, year-to-date, and life-to-date.

Inventory

For those contractors who carry a significant level of inventory, a module to monitor this activity is valuable. An inventory module should allow the user to track inventory by *date received*, *vendor*, and *cost*, as well as the distribution of inventory by *cost* and *job*. Many such modules can evaluate average unit costs, effective re-order points, and the most economical order quantities.

The inventory module is updated upon receipt of invoices using purchase order data and actual pricing information. This system provides cost data to the job cost and general ledger modules. The system should handle back orders and multiple price levels for various quantities of goods. Ideally, the inventory module will also update the billing module for time and material projects.

Reports

1. **Inventory Status**—listing stock levels, location, items on order, and back-ordered items.
2. **Monthly Activity**—listing receipts and issuance of inventory items by dates, vendors, prices, and jobs.
3. **Inventory Price List**—listing the purchase price and selling price for each item.
4. **Posted Costs**—listing all inventory items and associated costs issued to each project on a monthly basis.

Summary

Computers offer tremendous advantages to most contracting firms. When properly integrated, they can enhance internal control and profitability, while providing managers with more time to devote to other aspects of the business. Successful computerization requires an investment—of time and money—for researching the options for hardware and software and, possibly, restructuring operations for efficient automation.

APPENDICES

Table of Contents

INCOME STATEMENT

	CATEGORY NUMBER	CLASSI-FICATION NUMBER	TITLE NUMBER	GENERAL LEDGER NUMBER
I. CONTRACT REVENUE	40000			
A. DESIGN FEES				41100
B. BILLING ON WORK IN PROCESS				41300
C. PERCENT COMPLETE ADJUSTMENT				41400
II. COST OF SALES	50000			
A. DIRECT COSTS		51000		
1. Labor				51200
2. Material				51300
3. Subcontract				51400
4. Equipment				51500
5. Miscellaneous				51600
C. INDIRECT COSTS		53000		
1. Field Employee Expenses			53100	
a. Holidays				53110
b. Insurance – Health				53120
c. Insurance – Workers Comp.				53130
d. Payroll Taxes				53140
e. Retirement				53150
f. Salaries – Bonus				53160
g. Salaries – Vacation, Sick, Other				53170
h. Salaries – Supervisors				53180
2. Field Equipment Expenses			53200	
a. Depreciation				53210
b. Rental				53220
c. Repairs				53230
3. Field Insurance Expenses			53400	
a. Insurance-General Liability				53410
b. Builders' Risk				53420
4. Field Vehicle Expense			53300	
a. Depreciation				53310
b. Fuel & Oil				53320
c. Insurance				53330
d. Repairs & Maintenance				
5. Warranty				53500
6. Other Indirect			53600	
a. Miscellaneous				53610
b. Professional Fees				53620
c. Small Tools & Supplies				53630
d. Travel				53640

Figure A.1

	60000	61000		
III. OPERATING EXPENSES	60000			
A. GENERAL AND ADMINISTRATIVE		61000		
1. Business Insurance			61100	
a. Building				61110
b. Liability & Office Equipment				61120
c. Other				61130
2. Occupancy Expenses			61140	
a. Rent				61150
b. Repairs and Maintenance				61160
c. Taxes				61170
d. Telephone				61180
e. Utilities				61190
3. Office Employee Expenses			61200	
a. Continuing Education				61210
b. Insurance – Disability				61220
c. Insurance – Health				61230
d. Insurance – Workers Comp.				61240
e. Payroll Taxes				61250
f. Retirement				61260
g. Salaries – Bonus				61270
h. Salaries				61280
i. Social Activities				61290
4. Office Equipment Expenses			61400	
a. Depreciation				61410
b. Rental				61420
c. Repairs				61430
5. Office Supplies				61490
6. Office Vehicle Expense			61500	
a. Depreciation				61510
b. Fuel & Oil				61520
c. Insurance				61530
d. Repairs & Maintenance				61540
7. Officers Compensation			61600	
a. Insurance-Disabilit1y				61610
b. Insurance-Life				61620
c. Salaries-Bonuses				61630
d. Salaries				61640
e. Retirement				61650
8. Professional Fees			61700	
a. Accounting				61710
b. Legal				61720
c. Other				61730
9. Sales & Marketing			61800	
a. Advertising				61810
b. Commissions				61820
c. Literature				61830
d. Promotions				61840

Figure A.1 (continued)

10. Travel & Entertainment			61850	
a. Meals				61860
b. Travel				61870
11. Other Expenses			61900	
a. Bad Debts				61910
b. Branch Office Expense				61920
c. Classified Advertising				61930
d. Contributions				61940
e. Dues & Subscriptions				61950
f. Fines & Penalties				61960
g. Licensing				61970
h. Miscellaneous				61980
i. Temporary Secretarial				61990
IV. OTHER	70000			
A. INTEREST EXPENSE		71000		
1. Line of Credit				71100
2. Equipment Loans				71200
3. Other				71300
B. INTEREST INCOME				72000
C. NON-OPERATING INCOME & EXPENSE (net)				73000
V. INCOME TAXES	80000			
A. FEDERAL				81000
B. STATE				82000

Figure A.1 (*continued*)

FINANCIAL STATEMENT ACCOUNTS
BALANCE SHEET

ACCOUNT	CATEGORY NUMBER	CLASSI-FICATION NUMBER	TITLE NUMBER	GENERAL LEDGER NUMBER
I. ASSETS	10000			
A. CURRENT		11000		
1. Cash and Equivalents			11100	
a. Disbursement – Computer				11110
b. Disbursement – Manual				11120
c. Disbursement-Payroll				11130
d. Money Market				11140
e. Petty Cash				11150
f. Short Term Investment				11160
2. Receivable			11300	
a. Notes Receivable-Current				11310
b. Accounts Receivable-Progress Billings				11320
c. Unbilled Receivable				11330
d. Officer Receivable				11340
e. Employee Receivable				11350
3. Inventories			11400	
a. Marketing Literature				11410
b. Office Supply				11420
c. Other Materials				11430
4. Prepaid Expenses			11500	
a. Deposits				11510
b. Insurance				11520
c. Miscellaneous				11530
d. Taxes				11540
B. PROPERTY PLANT & EQUIPMENT				
1. Fixed Assets		12000	12100	
a. Office Equipment				12110
b. Accum. Depreciation – Off. Equip.				12120
c. Construction Equipment				12130
d. Accum. Depreciation – Const. Equip.				12140
e. Vehicles				12150
f. Accum. Depreciation – Vehicles				12160
g. Leasehold Improvements				12170
h. Accum. Depreciation – Leasehold Imp.				12180

Figure A.1 (continued)

C. OTHER		13000		
1. Deferred Taxes				13100
2. Long Term Investments			13200	
a. Cash Surrender Value Officers Life				13210
b. Marketable Securities				13220
c. Real Estate				13230
II. LIABILITIES	20000			
A. CURRENT		21000		
1. Payables			21100	
a. Notes Payable–Current				21110
b. Trade Payable				21120
c. Overbillings				21130
2. Accrued Liabilities			21200	
a. Payroll				21210
b. Payroll Taxes				21220
c. Trade Payable				21230
d. Vacation Compensation				21240
e. Retirement Payable				21250
f. State Sales Tax				21260
3. Employee Payroll W/H			21300	
a. FICA W/H				21310
b. FED W/H				21320
c. STATE W/H				21330
d. Life Insurance W/H				21340
e. Retirement W/H				21350
4. Corporate Income Taxes Payable			21400	
a. Federal Income Tax				21410
b. State Income Tax				21420
B. LONG TERM		22000		
1. Notes Payable			22100	
a. Office Equipment				22110
b. Construction Equipment				22120
c. Vehicle				22130
d. Other				22140
2. Deferred Taxes			22200	
a. Federal Income				22210
b. State Income				22220
III. STOCKHOLDERS EQUITY	30000			
A. CAPITAL STOCK				31000
B. PAID IN CAPITAL				32000
C. RETAINED EARNINGS				33000

Figure A.1 (continued)

JOB COST ACCOUNTS

DIVISION	SECTION	ACCOUNT DESCRIPTION	MATERIAL 51191	LABOR 51192	SUB-CONTRACTOR 51193	EQUIPMENT 51194	MISCELLANEOUS 51195
51190		GENERAL REQUIREMENTS					
	51200	Field Supervision	51201	51202	51203	51204	51205
	51210	Design/Drafting	51211	51212	51213	51214	51215
	51220	Temporary Services	51221	51222	51223	51224	51225
	51230	Cleaning	51231	51232	51233	51234	51235
	51240	Misc. Gen Requirements	51241	51242	51243	51244	51245
51250		SITEWORK	51251	51252	51253	51254	51255
	51260	Demolition	51261	51262	51263	51264	51265
	51270	Earthwork/Sediment Control	51271	51272	51273	51274	51275
	51280	Caissons/Pilings	51281	51282	51283	51284	51285
	51290	Drainage	51291	51292	51293	51294	51295
	51300	Utilities	51301	51302	51303	51304	51305
	51310	Paving	51311	51312	51313	51314	51315
	51320	Curb/Gutter/Walks	51321	51322	51323	51324	51325
	51330	Site Lighting	51331	51332	51333	51334	51335
	51340	Signage	51341	51342	51343	51344	51345
	51350	Seeding/Landscaping	51351	51352	51353	51354	51355
	51360	Misc. Site Improvements	51361	51362	51363	51364	51365
51370		CONCRETE	51371	51372	51373	51374	51375
	51380	Footings	51381	51382	51383	51384	51385
	51390	Slabs	51391	51392	51393	51394	51395
	51400	Foundation/Substructure	51401	51402	51403	51404	51405
	51410	Precast Concrete	51411	51412	51413	51414	51415
51420		MASONRY	51421	51422	51423	51424	51425
	51430	Block Work	51431	51432	51433	51434	51435
	51440	Veneer	51441	51442	51443	51444	51445
51450		METAL	51451	51452	51453	51454	51455
	51460	Structural Steel	51461	51462	51463	51464	51465
	51470	Pre-Engineered Steel	51471	51472	51473	51474	51475
	51480	Misc. & Ornamental Metals	51481	51482	51483	51484	51485
	51490	Fasteners/Rough Hardware	51491	51492	51493	51494	51495
51500		CARPENTRY	51501	51502	51503	51504	51505
	51510	Rough	51511	51512	51513	51514	51515
	51520	Trim/Stairs	51521	51522	51523	51524	51525
	51530	Cabinetry	51531	51532	51533	51534	51535
51540		MOISTURE/THERMAL/PROTECTION	51541	51542	51543	51544	51545
	51550	Termite/Water/Dampproofing	51551	51552	51553	51554	51555
	51560	Insulation/Fireproofing	51561	51562	51563	51564	51565
	51570	Roofing/Sheet Metal/Access	51571	51572	51573	51574	51575
	51580	Siding/Exterior Trim	51581	51582	51583	51584	51585

Figure A.1 (continued)

DIVISION	SECTION	ACCOUNT DESCRIPTION	MATERIAL	LABOR	SUB-CONTRACTOR	EQUIPMENT	MISCELLANEOUS
51590		DOORS/WINDOWS/GLAZING	51591	51592	51593	51594	51595
	51600	Metal/Specialty Doors	51601	51602	51603	51604	51605
	51610	Windows	51611	51612	51613	51614	51615
	51620	Glazing Systems	51621	51622	51623	51624	51625
51630		FINISHES	51631	51632	51633	51634	51635
	51640	Drywall/Steel Studs	51641	51642	51643	51644	51645
	51650	Flooring/Ceramic Tile	51651	51652	51653	51654	51655
	51660	Acoustical Treatments	51661	51662	51663	51664	51665
	51670	Painting/Wallcoverings	51671	51672	51673	51674	51675
51680		SPECIALTIES	51681	51682	51683	51684	51685
51690		ARCHITECTURAL EQUIPMENT	51691	51692	51693	51694	51695
	51700	Appliances	51701	51702	51703	51704	51705
	51710	Other	51711	51712	51713	51714	51715
51720		FURNISHINGS	51721	51722	51723	51724	51725
51730		SPECIAL CONSTRUCTION	51731	51732	51733	51734	51735
51740		CONVEYING SYSTEMS	51741	51742	51743	51744	51745
51750		MECHANICAL	51751	51752	51753	51754	51755
	51760	Plumbing	51761	51762	51763	51764	51765
	51770	HVAC	51771	51772	51773	51774	51775
51780		ELECTRICAL	51781	51782	51783	51784	51785
	51790	Service/Gear/Wiring	51791	51792	51793	51794	51795
	51800	Specialty Systems/Fixtures	51801	51802	51803	51804	51805

Figure A.1 (*continued*)

FINANCIAL ACCOUNTING

I. POLICY

The firm shall maintain an appropriate accounting system and internal controls required to produce accurate financial statements and to safeguard corporate assets.

II. PROCEDURES

A. GENERAL

1. The president and controller shall ensure the firm's accounting software satisfies all its needs as efficiently as possible, and is updated or customized as required.

2. The controller shall implement the procedures reflected on the firm's accounting flow chart. Minor adjustments may be made if needed to more effectively accomplish the intent of the flow chart.

3. The controller shall establish and maintain a well organized filing system for all financial records. Such records shall be retained in accordance with the Documentation Policy.

4. For financial reporting, the firm shall use the accrual and percentage-of-completion method of recognizing revenue. For tax purposes only, the completed contract and cash methods will be used.

B. OPERATIONS

1. Labor
   a. The company labor force is scheduled for the following week at the weekly management meeting. Any changes in the schedule will be coordinated by the supervisors.

   b. Time sheets are completed on a daily basis by all hourly employees. Each supervisor is responsible for checking the job cost coding and approving the hours worked. Each hourly employee is responsible for submitting his time sheet for the previous week to the accounting assistant by 5:00 p.m. on the first working day of the week. If the time sheet is submitted late, the employee risks missing payroll for that week. Paychecks are prepared on Tuesday and distributed on Wednesday.

2. Materials
   a. Purchase orders are to be prepared for all material purchases in accordance with the Purchasing Policy.

Figure A.2

Suppliers are responsible for submitting their invoices to accounting on a timely basis.

    b. When a supervisor places the material order, he will determine the time and place of delivery.

  3. Subcontractors
    a. When the plans are available from the architect, they should be distributed to the various subcontractors for pricing, in accordance with the Purchasing Policy.

    b. Subcontract Agreements must be completed whenever a subcontractor is providing services. There should be certificates of appropriate insurance on file for each subcontractor before the subcontractor begins work. Each subcontractor is responsible for submitting his invoices on a timely basis.

C. MONTHLY PROCESSING

  1. Payroll
    a. The accountant will separate each time sheet into regular and overtime hours (hours <u>worked</u> in excess of 40 and not including vacation, holiday, or sick leave hours). In addition, the total hours should be distributed to the various job cost and/or indirect account codes. All applicable accounts are listed on the back of each time sheet for reference by field employees. The individual cost codings and hour distribution for each employee are entered into the computer.

    b. The computer will calculate gross pay and all deductions, arriving at net pay. Payroll checks are then printed and all job cost and general ledger files are updated. An employee register is updated regularly, listing wages or salary, raises, bonuses, and benefits for each employee.

    c. Checks are given to the president for signature. When returned, the yellow copy is maintained for internal control.

    d. Payroll checks should typically be available to employees by Wednesday after 3:30 p.m.

    e. Time sheets are filed weekly by month. The various computer reports are filed by report for each week.

  2. Billing
    a. The accounting assistant meets regularly with each supervisor to determine if a draw request can be sent for their job. After determining if a billing can be made, the draw request form should be completed. The total billing should then be recorded in the sales journal at the gross amount. Retention, if applicable, should not be deducted

Figure A.2 (continued)

when recording the billing in the sales journal.

b. The billing request should be mailed promptly with a copy retained to be filed in the job book.

c. The receivables report should be completed on a weekly basis for all open receivables and forwarded to the president for review.

d. When any receivable reaches past due status, the controller should follow up to determine if there are any problems. Action should be taken very early in the case of any overdue receivable, in accordance with the Credit Policy.

3. Cash Receipts
    a. When owner payments are received, they should be immediately restrictively endorsed with the company endorsement stamp.

    b. When the bank deposit is prepared, the accounting assistant should make copies of each check and deposit slip. A copy of the deposit slip should be attached to the check and filed with the cash receipts journal.

    c. A cash receipts journal should be maintained for each cash account and immediately updated when deposits are made.

    d. At least monthly, the controller should prepare a week-by-week cash flow forecast for the next 4-week period. Any excess cash can then be deposited in a money market fund to earn the highest yield.

4. Accounts Payable
    a. The secretary will deliver invoices received by mail to the accounting assistant, who will prepare an accounts payable voucher for invoices to be processed as follows:
        1. Material invoices should be matched to purchase orders and delivery tickets.
        2. Subcontractor invoices should be checked against subcontract agreements.
        3. Verify extension and footings of invoice.
        4. Accrue any sales tax not billed on invoice.
        5. Complete voucher form.

    b. When all invoices and vouchers are processed, they are forwarded to the appropriate managers for approval.

    c. After managers approve the invoices, they are entered into the computer system by vendor.

    d. The computer will update job cost, general ledger, and vendor files as each invoice is entered. After all invoices are entered, the accounts payable clerk will verify the dollar amount of the invoices by running an adding machine tape of each invoice amount and comparing this total to the

Figure A.2 (continued)

computer control total.

   e. Accounts payable vouchers are filed by due date by vendor.

  5. Cash Disbursements
    a. The accounting assistant will select invoices by due date when invoices become due, and review the list with the controller as necessary.

    b. The computer will print checks and update all general ledger and vendor files. For each check printed, the approved vouchers supporting the check should be attached to the checks.

    c. The checks and approved vouchers are then submitted to the president for approval. After the checks are signed, they are distributed to the payees.

    d. The pink copy of the check is matched with the vouchers and filed by vendor. The yellow copy is maintained for numerical control.

D. MONTH/YEAR END

  1. Payroll Tax
    a. Payroll tax deposits must be made at a Federal Reserve Depository within 3 banking days once a $3,000 liability is attained. In order to avoid any penalties, it is suggested these deposits be made every week.

    b. State withholding of income taxes must be remitted to the state within 15 days following the end of a month. An exception is made for the months of March, June, September, and December. For these months, the business is allowed until the end of the following month to make the remittance.

    c. On a quarterly basis the business is required to file Federal Form 941 reporting the liability for federal and FICA tax withholding. This report is due within 30 days of the end of the quarter.

    d. On a quarterly basis, the business is required to remit federal unemployment taxes calculated by applying the effective rate on the first $7,000 of each employee's wages. When the year to date tax liability exceeds $100, the tax must be deposited through the federal reserve system. This deposit is due within 30 days of the end of the quarter.

    e. On a quarterly basis, the business is required to file the state unemployment tax return and remit the tax with the return. The tax is calculated by applying the effective rate on the first $7,000 of each employee's wages. Deposits and tax returns are due within 30 days of the end of the quarter.

Figure A.2 (continued)

f. At the end of January of each year, the business is required
    to file federal form 940 reporting the annual liability for
    federal unemployment tax and prepare form W-2 distributing
    copies to each employee.

2. Job Cost
    a. The preliminary job cost report is printed for all jobs in
        progress. The president will then review the budget for each
        job and update the budget as necessary in accordance with
        current proposals and actual costs anticipated.

    b. The accounting assistant inputs all budget adjustments.

    c. The final job cost reports are printed for use in the
        percent complete adjustment and review by field supervisors.

3. Monthly Financial
    a. The following standard journal entries are prepared to
        adjust the accounts for monthly closing.
        #1 Cash receipts
        #2 Reverse of previous month's accruals
        #3 Accrued payroll
        #4 Employee payroll taxes (FICA, Federal & State
            Unemployment)
        #5 Billings
        #6 Amortization of prepaid insurance (Auto, General
            Liability, Worker's Compensation)
        #7 Accrual of retirement plan expense
        #8 Depreciation expense
        #9 Void checks
        #10 Record interest revenue
        #11 Percent complete adjustment, Cost of Sales entry
        #12 Tax entry
        There may be miscellaneous adjustments from month to month.
        Any entries that affect job cost accounts must be entered
        before printing final job cost reports.

    b. After the first ten standard journal entries and any
        miscellaneous entries are complete, they should be entered
        into the computer. The computer will post the entries to job
        cost and general ledger files.

    c. After the final job cost reports are printed, the controller
        prepares the gross profit report by job to adjust the
        billings to revenue earned. After completing this report,
        journal entry #11 can be completed and input.

    d. After printing preliminary financial statements, the
        controller should thoroughly review the detailed balance
        sheet and income statement for any errors. After entering
        any adjustments, entry #12 for income tax expense should be
        completed and entered. Final statements can then be printed.

    e. After all necessary reports are printed, the month end

Figure A.2 (continued)

closing can be completed by the computer. This function
updates all income and expense year to date balances and
zeros the current month amounts. Retained earnings is
updated for the current month's net income.

    f.   The controller completes the following managerial reports:
        1.   Project Status Report
        2.   Balance Sheet Ratios
        3.   Income Statement Ratios

4.   Year End Financial
    a.   The controller should communicate with the auditor in
       advance of the year end date to tentatively schedule the
       review.

    b.   The final detailed Income Statement and Balance Sheet should
       be printed to be given to the auditor for the review.

    c.   All final schedules substantiating the Balance Sheet
       accounts should be completed for use in the review.

    d.   The controller will meet with the auditor during the review
       process to answer any questions or provide additional
       information.

    e.   Any final adjustments as a result of the review should be
       entered into the general ledger.

5.   Income Tax
    a.   Federal Estimated Taxes are due on December 15, February 15,
       May 15, and August 15. Generally, the estimated tax will
       equal the tax liability from the preceding year end. The
       estimated tax must be paid through the Federal Depository
       system.

    b.   The same rules apply for state estimated tax except that the
       deposit should be sent to the Comptroller of the Treasury.

Figure A.2 (continued)

Appendix A

<u>PURCHASING POLICIES AND PROCEDURES</u>

I. POLICY

   A. The firm shall establish purchasing controls designed to ensure
      that the best prices are being obtained, that vendor bills match
      quotes, and that data for accurate job cost control is generated.

II. PROCEDURES

   A. Materials/Supplies

      1. Accounting Assistant

         a. The accounting assistant will maintain a log with the name
             of the person receiving purchase orders, the purchase order
             number(s), and the date released. As each purchase order is
             returned, it should be checked by the accounting assistant
             for completeness and recorded in the log. If information is
             incomplete or inaccurate, the purchase order should be
             returned to the issuer for correction. Copies of the
             purchase order should be distributed as follows: original
             copy filed in appropriate job book if job-related; yellow
             copy filed for accounting purposes; and pink copy given to
             supervisor if job-related.

      2. Managers

         a. All material purchases must have a written purchase order
             issued, including purchases of materials for jobs, small
             tools and supplies, equipment rental, warranty, truck
             maintenance, and office supplies. Each purchase should be
             inclusive of all materials needed for any given phase, with
             small subsequent purchases minimized. In all cases, a key
             rule is that one Purchase Order <u>must</u> match one invoice.
             Standard material package forms should be used when
             possible, double checking as required to avoid errors.
             Material packages should be sized to allow rapid use with
             minimum storage time.

         b. Upon completion, the purchase order should be priced by at
             least two vendors, preferably three on larger orders. When
             the manager has received all price quotes, he/she should do
             the following:
             -- analyze them for comparability, select the best vendor,
                and fill out the purchase order form with the price,
                vendor, date, job name, job cost account, and
                description of materials
             -- submit the purchase order, along with the price quotes
                to the accounting assistant for approval by the
                president prior to ordering if the amount exceeds $500

Figure A.3

256

             (request help in securing prices from the accounting
             assistant when needed)

           -- proceed with the purchase and submit the purchase order
             directly to the accounting assistant within one week of
             ordering if less than $500 (no advanced approval
             required in this case)

           -- communicate the purchase order number to the selected
             vendor for recording on the invoice

    c.   Purchases of all material packages for different phases
        shall be done on one single purchase order when the package
        is to be delivered in one shipment. When multiple shipments
        over time are involved, separate purchase orders shall be
        prepared simultaneously for each individual delivery. The
        one exception to this shall be concrete and gravel, which
        can be submitted on one Purchase Order to include all
        deliveries required for the project by separate phase
        (footings, slabs, foundation).

    d.   Variance purchase orders are to be used for subsequent
        purchases that exceed the original estimate and original
        purchase order in any given division. An explanation shall
        be provided, e.g. changes, theft, underestimating. All
        variance Purchase Orders, regardless of the value, are to
        be submitted for approval in advance of ordering.

    e.   Purchase Orders may be adjusted for quantity if it has not
        yet been submitted to the accounting assistant. Once the
        Purchase Order has been given to the accounting assistant,
        no such changes can be made. When adjustments are made in
        quantities, the supervisor is responsible to secure and
        record the adjusted price.

    f.   The supervisor is responsible for confirming the
        availability of all materials with the supplier. In lieu of
        back orders on any given Purchase Order, a separate
        Purchase Order shall be prepared for materials not
        immediately available.

3. Material Invoices

    a.   As suppliers submit invoices, the accounting assistant will
        attach the appropriate voucher and submit them to the
        supervisor by Tuesday afternoon. The supervisor will review
        the bill and ensure it is in line with the purchase order,
        that all quantities were actually delivered, and all
        materials were of good quality. If so, the supervisor
        initials the voucher and returns it to the accounting
        assistant by Thursday afternoon. All invoices are to remain
        in the central offices.

Figure A.3 (continued)

4. Subcontractor Invoices

a. When subcontractors submit bills, the accounting assistant will attach the appropriate voucher and submit them to the supervisor by Tuesday afternoons. The supervisor will then review the bill and ensure it is in compliance with the draw schedule and does not result in an overpayment.

b. If acceptable, the supervisor will initial the voucher and return it to the accounting assistant by Thursday afternoon of the same week. The invoices are not to be removed from the central offices. Final payments will be made to subcontractors only when the fully signed Quality Control Checklist has been submitted by the supervisor. The accounting assistant should attach the completed checklist to final payment checks for the president's review prior to signing.

5. Improvements

a. Each tenant improvement estimate should be in the 16 Division format and job costed as individual projects. Markup on tenant improvements should be no less than 25%, and possibly more for smaller contracts. Each is to be individually reviewed with the president.

b. Every attempt must be made by the supervisors and accounting assistant to isolate tenant improvements direct costs from original contract costs. All extras from subcontractors for tenant improvements should be clearly described on their bills. All such bills must be supported by proposals that are secured and inserted into the job notebook by the supervisor.

c. Lumber and trim required under the original contract shall be ordered on individual purchase orders for each unit. Any extra materials required above and beyond the original contract quantity take-offs shall be ordered for each tenant on separate purchase orders designated for tenant improvements.

6. Other

a. Personnel may make substantial personal purchases through the company when significant savings are available. All personal purchase orders, regardless of the amount, should be submitted to the president for approval prior to placing the order. Payment to the company should be made when the material is ordered.

B. Subcontractors

Figure A.3 (*continued*)

1. Responsibility

    a. The project supervisors are typically responsible for securing subcontractor bids for their respective projects. However, depending on the circumstances, a portion of this responsibility may be assumed by the president, design director, or administrative assistant.

    b. The design director will be responsible for maintaining a log of approved subcontractors with a proven track record and good references. A qualification form is to be available for gathering such information on new firms.

2. Bid Documents

    a. The responsible party should typically pursue a minimum of three bids for each phase of construction. Each subcontractor is to be provided identical documents, including the plans, specifications, Quality Control Checklist, Addendum A (if applicable), Safety Checklist, and Bid Form if appropriate. Addendum A is a specific description of all work and materials to be provided by the subcontractor.

    b. A specific due date for bid receipt should be outlined either verbally or in writing. Bids should always be submitted in writing.

    c. Upon receipt of bids, the secretary should file the original in the construction notebook, and submit a copy to the soliciting manager.

3. Evaluation

    a. The responsible manager should evaluate the bids in terms of completeness, promptness, price, and qualifications of the bidders. A recommendation for award is then made to the president.

Figure A.3 (continued)

FIELD DOCUMENTATION

I. POLICY

    A. The firm shall establish a control system to ensure all
       field-related policies, procedures, and systems are observed,
       with adequate communication to the office.

II. PROCEDURES

    A. Accounting Assistant

       1. The accounting assistant shall be responsible for collecting
           and processing all field paperwork. The attached Exception
           Report, indicating all variances, is to be maintained
           continuously and submitted to the president during the week
           following the end of each month.

    B. Secretary

       1. The secretary shall be responsible for filing all forms and
           records in the appropriate locations, and providing
           secretarial support for processing such forms.

    C. Supervisors

       1. The supervisor of each project shall be responsible to ensure
           a series of forms are submitted to the accounting assistant
           in a timely fashion, and in accordance with the applicable
           procedures for each form. If any delay or difficulty is
           anticipated in meeting the deadlines, it should be
           communicated to the president before the deadline is passed.

       2. The forms and associated procedures are as follows:

           a. Change Orders: Completed and signed prior to execution of
               any changes, in accordance with the Changes Policy.

           b. Computer Schedule: Computer printout prepared within 2
               weeks of the start of construction, and updated within 3
               days when requested by the president, or as needed by the
               supervisor.

           c. Daily Field Reports: Submitted preferably by each Friday,
               but no later than Monday morning of the following week.

           d. Draw Requests: Return to accounting assistant with all
               adjustments within two days of receipt. Accounting
               assistant reviews with president prior to preparation.

           e. Injury Report: Submitted within 1 day of any accident as
               outlined in the Field Safety Policy.

           f. Invoices: Picked up Tuesday afternoons and returned with
               approvals or comments by Thursday afternoons.

           g. Management Checklist: Submitted to the office by the last
               day of each month.

Figure A.4

h. Miscellaneous Correspondence: letters and memos issued promptly to all affected individuals (subcontractors, president, controller, suppliers, etc.) in any situation that may result in misunderstandings, contractual conflict, safety violations, etc. Consult with the president if any doubt exists as to the potential for a problem becoming serious.

i. Progress Meeting Minutes: submitted within 1 day of each progress meeting. Typed by accounting assistant and mailed to all subcontractors. Give one copy to the president.

j. Project Completion Party: Submit a list of all individuals to be invited to the Project Completion Party 6 weeks prior to the completion of the project.

k. Purchase Orders: submitted in accordance with the Purchasing Policy prior to ordering materials.

l. Quality Control Checklist: one copy signed by the subcontractor's foreman and submitted within the week the respective work is started. One final copy signed by the supervisor and submitted within the week the work is completed and prior to final payment to the subcontractor.

m. Safety Checklist: submitted to the office by the last day of each month.

n. Subcontract Agreement: completed and signed by both parties prior to the subcontractor's start of work.

o. Subcontractor Rejection Letter: mailed to all rejected bidders immediately after award of contracts. Verbal contact is also encouraged.

p. Substantial Completion Form: completed at walk-through inspection prior to occupancy by any owner or tenant. Financial adjustments put in writing and signed.

q. Tenant Contracts: Prepare estimates and specifications, and submit to president for review. Submit adjusted details to accounting assistant for contract preparation. Ensure contract signed by both parties prior to starting work.

r. Time Slips: Supervisor and field assistants' time slips submitted by Monday afternoons. Field assistants' time slips to be initialed each day by the supervisor.

s. Two-Week Schedule: Submitted to the office by Friday morning of each week.

Figure A.4 (continued)

EASTWAY CONTRACTORS
ESTIMATE SUMMARY

PROJECT: _____ LOCATION: _____

ARCHITECT: _____ ESTIMATOR: _____

SQUARE FEET: _____ DATE: _____

DIV.	DESCRIPTION	MATERIAL	LABOR	SUBCONT.	EQUIP.	MISC.	SUBTOTAL	TOTAL
1.0	**GENERAL REQUIREMENTS**							
	A. Field Supervisor							
	B. Design/Drafting							
	C. Temporary Services							
	1. Fencing, Barricades							
	2. Office/Telephone							
	3. Power/Lighting/Water							
	4. Rental Equipment							
	5. Signage Areas							
	6. Storage Areas							
	7. Temporary Roadways							
	8. Toilets							
	9. Winter Protection							
	SUBTOTAL							
	D. Cleaning							
	1. Trash Clean-up/Removal							
	2. Final Cleaning							
	SUBTOTAL							
	E. Miscellaneous Gen. Req.							
	1. Bonds							
	2. Permits/Inspection Fees							
	3. Photographs							
	4. Sales Taxes							
	5. Testing							
	SUBTOTAL							

Figure A.5

DIV.	DESCRIPTION	MATERIAL	LABOR	SUBCONT.	EQUIP.	MISC.	SUBTOTAL	TOTAL
2.0	**SITEWORK**							
	A. Demolition							
	B. Earthwork/Sediment Control							
	C. Caissons/Pilings							
	D. Drainage							
	E. Utilities							
	F. Paving							
	G. Curb/Gutter/Walks							
	H. Site Lighting							
	I. Signage							
	J. Seeding/Landscaping							
	K. Misc. Site Improvements							
3.0	**CONCRETE**							
	A. Footings							
	B. Slabs							
	C. Foundation/Superstructure							
	D. Precast							
4.0	**MASONRY**							
	A. Blockwork							
	B. Veneer							
5.0	**METAL**							
	A. Structural Steel							
	B. Pre-Engineered Steel							
	D. Fasteners/Rough Hardware							
	C. Misc. & Ornamental Metals							

Figure A.5 (continued)

DIV.	DESCRIPTION	MATERIAL	LABOR	SUBCONT.	EQUIP.	MISC.	SUBTOTAL	TOTAL
6.0	**CARPENTRY**							
	A. Rough							
	B. Trim/Stairs							
	C. Cabinetry							
7.0	**MOISTURE/THERMAL PROTECT.**							
	A. Termite/Water/Dampproofing							
	B. Insulation/Fireproofing							
	C. Roofing/Sheet Metal/Access.							
	D. Siding/Exterior Trim							
8.0	**DOORS/WINDOWS/GLAZING**							
	A. Metal/Specialty Doors							
	B. Windows							
	C. Glazing Systems							
9.0	**FINISHES**							
	A. Drywall/Steel Studs							
	B. Flooring/Ceramic Tile							
	C. Acoustical Treatments							
	D. Painting/Wallcoverings							
10.0	**SPECIALTIES**							
11.0	**ARCHITECTURAL EQUIPMENT**							
	A. Appliances							
	B. Other							
12.0	**FURNISHINGS**							
13.0	**SPECIAL CONSTRUCTION**							

Figure A.5 (continued)

DIV.	DESCRIPTION	MATERIAL	LABOR	SUBCONT.	EQUIP.	MISC.	SUBTOTAL	TOTAL
14.0	**CONVEYING SYSTEMS**							
15.0	**MECHANICAL**							
	A. Plumbing/Sprinkler							
	B. HVAC							
16.0	**ELECTRICAL**							
	A. Service/Gear/Wiring							
	B. Specialty Systems/Fixtures							

INDIRECT CONSTRUCTION COSTS

- Material _____ x _____ % =
- Labor _____ x _____ % =
- Subc. _____ x _____ % =
- Equip. _____ x _____ % =
- Misc. _____ x _____ % = _____

      Total Indirect Costs

DIRECT CONSTRUCTION COSTS $

INDIRECT CONSTRUCTION COSTS

GENERAL AND ADMINISTRATIVE EXPENSES ____%

SUBTOTAL

PROFIT ____%

TOTAL BID $

Figure A.5 (continued)

BUSINESS PLAN
CHECKLIST

I. **Corporate History & Purpose**

   A. Date, location, and players in initial business formation
   B. Primary business purpose and motivation
   C. Critical development dates and events
   D. Types of products/services
   E. General market conditions/area
   F. Customer base
   G. Opportunities/threats
   H. Long term general goals
   I. Sales/Financial history

II. **Strengths and Weaknesses**

   A. Personnel
   B. Quality
   C. Control systems
   D. Leadership
   E. Public relations
   F. Work flow
   G. Financial

III. **Goals and Strategies**

   A. Strategic
      1. Business image
      2. Relationships with associates
      3. Government relations

   B. Coordinative
      1. Management style
      2. Responsibility/authority
      3. General work flow
      4. Facilities
      5. Standards of performance

   C. Operational
      1. Sales/profits
      2. Productivity
      3. Scheduling
      4. Purchasing

IV. **Marketing Plan Summary**

   A. National, regional, and local economic conditions
   B. Competition
   C. Sales and profit goals -- short and long term
   D. Current and anticipated contracts/customers
   E. Marketing strategies
   F. Marketing budget
   G. Diversification opportunities

V. **Organizational Plan**

   A. Current and planned organizational structure
   B. Managerial style (level of formality, centralization, entrepreneurship, competition)
   C. Individual/departmental relationships
   D. Current and planned work flow
   E. Responsibility/Authority
   F. Adjustments to policies, procedures, job descriptions

VI. **Financial Plan**

   A. Operating budget
   B. Cash flow budget
   C. Pro Forma
   D. Adjustments in accounting procedures, assets, debt, taxation
   E. Corrections to negative financial trends
   F. Financial goals for each manager/department
   G. Financial structure goals

Figure B.1

EASTWAY CONTRACTORS

1988 BUSINESS PLAN

TABLE OF CONTENTS

Figure B.2

Corporate History and Purpose

Eastway Contractors was established by John Eastway in 1980 as a small tenant improvement contracting firm with offices located in Washington, D.C. As the competitive-bid tenant improvement market in Washington diminished during the recession of 1980-82, the firm moved quickly into the light commercial construction market and expanded its area of operation to include several peripheral locations in Maryland and Virginia. The light construction industry in this region has been very strong since 1983, allowing Eastway to steadily increase its annual revenues. While the U.S. economy is expected to experience a slow-down during the next several years, and the full impact of current tax reform is yet unknown, it is expected that the Washington market will remain relatively stable.

Since 1984, Eastway has focused primarily on the design/build approach (negotiated contract/comprehensive services) to its projects, as opposed to the traditional approach (competitive-bid/limited services) more typical in the industry. The result has been a more stable level of growth in revenues, greater profitability, and a far more secure market position. In order to maintain this positive direction, John Eastway has charged the firm with the primary purpose of delivering high quality construction services on progressively larger projects, while earning above-average returns on sales and equity.

Strengths and Weaknesses

The top management of Eastway recognizes certain strengths and weaknesses within the firm that directly affect its ability to accomplish its primary purpose. Important strengths include:

    *   highly qualified personnel
    *   excellent quality control in the field
    *   generally strong internal control systems
    *   excellent service to customers

Significant weaknesses include:

    *   lack of a strong local image and the associated business
         contacts
    *   occasional problems with employee morale and productivity
    *   inefficiency in dealing with many complex regulatory
         standards (codes, permits, environmental matters,
         inspections, etc.)
    *   poor purchasing procedures in the field, sometimes causing
         excessive costs and confusion with paperwork in the office
    *   inconsistent work flow and/or scheduling in the field,
         causing occasional shortages or excesses of labor and
         staff capacity
    *   poor cash flow control, requiring excessive use of short
         term credit to cover all direct costs and operating
         expenses

Figure B.2 (continued)

<u>Goals and Strategies</u>

Eastway has established a number of goals intended to address the weaknesses within the firm and to facilitate the realization of the firm's primary purpose. These goals fall into three basic categories: strategic -- concerning the firm's activities relative to its external business environment; coordinative -- relating to the coordination of activities between organizational levels and functions; and operational -- concerning specific, short term, and measurable goals at the operations level.

Strategic Goals

1.   Goal: Establish a stronger local business image with the public in general and important business contacts in particular.

      Strategy: See Marketing Plan Summary

2.   Goal: Acquire greater control over the timing of new projects, providing a more even work flow for the field and office staffs.

      Strategy: See Marketing Plan Summary

3.   Goal: Establish improved means of communication with regulatory agencies, as well as improved systems to address all government requirements.

      Strategy:
      1.   Key office and field managers will personally meet the regulators with whom they have regular dealings to develop a better relationship. Managers will be encouraged to send notes thanking regulators whenever applicable.

      2.   Managers will attend a seminar on the subject of negotiation to learn valuable techniques in working with regulators and other business associates.

      3.   The administrative assistant will prepare regulatory notebooks for each of the firm's managers who must deal with these issues. Each notebook will include all applicable regulatory documents and references (permit forms, codes, fee structures, etc.), as well as compliance checklists. All regulatory questions or problems will be handled by the administrative assistant.

Coordinative Goals

1.   Goal: Encourage more entrepreneurship and team spirit in an effort to improve morale and productivity.

Figure B.2 (continued)

Strategy:
1. The president will design a more performance-oriented system for establishing the level of bonuses to be awarded to various individuals.

2. Job descriptions will be reviewed by the president to determine if various responsibilities may be further delegated, with an eye toward more decentralization and self-management.

3. Organizational goals and strategies will be more fully communicated to all affected personnel, as well as their respective roles in accomplishing those goals.

4. Top management will be encouraged to pursue more casual, less rigid relationships with other individuals within the firm, and to offer positive reinforcement whenever possible.

2. Goal: Attain better control over field purchasing.

Strategy:
1. The president will review the existing purchasing system with the controller, accounting assistant, and field supervisors, and correct any current procedural inefficiencies.

2. The controller will establish a more comprehensive system of purchasing checks and balances. The president will ensure that the annual performance evaluations and the associated bonuses or merit increases comply with the purchasing system.

3. The controller will communicate the purchasing procedures to vendors on a more formal level. It should be made clear to the vendors that they must comply with purchasing procedures.

Operational Goals

1. Goal: Increase sales volume to eight million dollars with a net profit before taxes in the range of 4% - 5%.

   Strategy: See Marketing Plan Summary and Financial Plan.

2. Goal: Acquire greater control over cash flow and reduce the dependence on short term credit.

   Strategy: See Financial Plan.

3. Goal: Attain greater control over field scheduling procedures.

Figure B.2 (continued)

Strategy:
1. The administrative assistant will enroll all field supervisors in a seminar on the subject of construction scheduling.

2. The vice president will conduct an internal review of CPM scheduling techniques and work more closely with individual supervisors in the development of their respective schedules.

3. The vice president will establish a more formal system of monitoring the updating of CPM schedules and ensuring proper consolidation and communication of all schedules to the affected vendors.

Marketing Plan Summary

The marketing plan prepared by John Eastway describes in detail the marketing goals and strategies that the firm will pursue in 1988. Important points applicable to this business plan are summarized as follows:

1. The national and regional construction market is generally positive, although there has been overbuilding in certain areas. Local construction activity is expected to be strong in 1988 except in office buildings. Retail and warehousing offer the greatest opportunities. Many owners are recognizing the time and cost saving benefits of the design/build approach to light commercial projects. Competition in the design/build sector is expected to be mild, allowing many opportunities for Eastway.

2. The financial plan projects sales revenues of eight million dollars with a net profit margin before taxes of 4%. The president intends to acquire at least 50% of the sales revenue through internally-generated projects (development partnerships). The attached Project Status Report identifies specific projects within the firm's current market area capable of accomplishing these two goals. In addition, the president is currently pursuing a number of other projects capable of yielding a similar or greater volume in 1989.

3. The president and vice president will become more involved in the local Chamber of Commerce, Rotary Club, and certain high profile community projects. Closer contacts will be made with local political figures, economic development officials, and commercial brokers. The design/build capabilities of the firm will be stressed.

Figure B.2 (continued)

4. A stronger and more consistent corporate image for service and quality will be developed through project signage, brochures, and advertising. Publicity will also be pursued through ground breakings and grand openings. A total of $47,000 is expected to be spent on marketing, including:

Advertising	$ 8,000
Commissions	30,000
Literature	2,000
Promotions	7,000
	$47,000

5. The president and vice president will attend a two-day seminar on selling construction services.

6. The firm will establish a small service crew capable of performing maintenance work, tenant improvements, and small remodeling projects. This will be an additional service offered to past and prospective customers, while allowing the firm to smooth out the peaks and valleys in its work flow. A promotional effort will be designed specifically for this new service.

Organizational Plan

The attached organizational charts indicate three departments: Project Management, Technical Services, and Office Administration. Each of these departments is shown as a whole, minimizing the negative psychological effect of rigid hierarchial relationships and departmental scaling. While there is significant horizontal differentiation between departments, vertical differentiation between individuals within each department is not emphasized, thereby creating a sense of equality and unity. Internal relationships are more clearly defined in job descriptions as well as the Policies and Procedures Manual.

The firm's projected organizational chart for 1988 reflects the addition of a vice president, one additional supervisor, and one additional field assistant. These changes are required due to the additional volume and new complexities involved in managing the firm. All job descriptions and internal control systems affected by the addition of a vice president will be properly adjusted in writing and communicated to the affected individuals.

The internal work flow of Eastway Contractors is diagrammed on the attached flow chart. While this flow chart indicates a somewhat fixed operational pattern, there are actually a number of alternative approaches to accomplishing many of the required tasks.

While responsibility and authority among individuals is clearly established through job descriptions and the policies and procedures, they are applied in an informal manner which focuses on individuals and relationships rather than positions and formal

Figure B.2 (continued)

rules. Stated more simply, a certain level of informality and an individual approach in accomplishing goals is tolerated, without absolute adherence to every aspect of all formal systems. As the firm continues to grow and deal with more complex projects, the level of this informality must be monitored by top management.

The firm has developed a series of policies and procedures which outline the means whereby information is managed and specific tasks are accomplished. These policies are generated in preliminary form by top management, with input and refinements encouraged from all participants. While individual managers do not have the authority to change policies, they often have the authority to modify procedures required to meet a specific policy pertaining to their activities. This authority is in the job descriptions, outlined with major modifications generally discussed in advance with the president or in the weekly management meeting.

Financial Plan

Operating Budget: The attached 1988 Budget Summary reflects the direct, indirect, and general and administrative costs the firm expects to incur at the projected revenue level of $8 million. These costs are projected with due respect to their fixed, variable, or semi-variable components. The budget projects:

a.   direct costs of $6.88 million, or 86% of sales
b.   total operating costs of $807,029, or 10.09% of sales
c.   operating profit of $312,971, or 3.91% of sales

These projections seem quite reliable, given historical operating cost data, and the Marketing Plan assumptions.

Cash Flow Budget: The firm has experienced fairly regular problems with cash flow and has depended excessively on its line of credit. Strategies to improve this situation include the following:

1.   The president will consult with the corporate attorney on improvements in contractual language pertaining to the timing of payments by customers. Customers and bankers will be more formally apprised of all collection terms.

2.   A receivables report, including action taken, will be prepared weekly by the accounting assistant and reviewed with the controller. The controller will take appropriate action weekly. Specific problems will be communicated weekly to the president.

3.   The annual cash flow projection, and all associated individual project cash flow projections, will be updated by the controller and reviewed with the president monthly.

The attached Annual Cash Flow Budget reflects a maximum ending

Figure B.2 (continued)

short term debt of $100,000 in May, and no such debt in 5 of the 12 months in 1988. This is a substantial improvement over 1987, but it will depend on careful scheduling, billing, and collection follow-up.

Pro Forma Statements: The attached pro forma balance sheet and income statement reflect the firm's actual performance in 1987 and that projected for 1988. The projected income figures are directly related to the annual budget for income and expenses. The balance sheet reflects the anticipated changes in current assets, fixed assets, current liabilities, long term debt, and stockholders' equity.

Accounting Methods: The firm currently uses the accrual basis of accounting both for financial and tax reporting. While income recognition for tax purposes is currently on the completed contract basis, the financial reporting system uses the percentage of completion method. However, management recognizes that in 1989, when it is expected revenues will exceed $10 million, the firm will be required to use the percentage of completion method for tax reporting. No other significant changes in accounting methods are anticipated.

Internal job costing techniques for tenant improvement contracts on the firm's retail and office projects have been a consistent source of confusion. The controller will establish improved means for collecting and reporting such cost data such that all tenant improvement costs are isolated from the original building contract.

Fixed Assets: The firm has maintained a policy of limiting all fixed assets in favor of renting equipment and utilizing subcontractors. If the tax laws are modified in favor of asset ownership, the firm will reconsider this position. The offices are leased from the president, John Eastway. In 1988, the firm plans to purchase only two new trucks and a few light field tools.

Debt: Each year the controller and president assemble a complete informational booklet for the firm and distribute it to major creditors and the firm's bonding company. This booklet includes all financial, managerial, and organizational documents of specific interest to creditors.

Eastway has developed excellent credit relationships with several banks. National Bank has established a line of credit in the amount of $400,000 for the firm's use in covering short term cash needs. It is the president's intent to increase the level of this line to the maximum level each year. Because the firm has limited its fixed assets, the need for long term debt is minimized.

Taxation: The president and controller will review several tax-related issues with the firm's tax accountant and attorney in order to explore methods of maximizing after-tax corporate earnings and income to the president. Specific areas of opportunity or concern include:

Figure B.2 (continued)

1. Treatment of the new tenant improvement and maintenance department as a separate corporation, in order to capitalize on the stepped corporate income tax rates.

2. Proper documentation of the need for high levels of working capital in order to avoid an accumulated earnings tax.

3. Consideration of the Sub S Corporation election in order to limit the taxation of further earnings to only the president's personal level, and to eliminate the issue of accumulated earnings.

4. Consideration of new "perks" for top management that can be paid for by the firm as deductible expenses. Specific possibilities include:

    a. corporate vehicles
    b. expanded travel and entertainment privileges
    c. a more favorable qualified pension plan
    d. a deferred compensation plan

5. Purchase of the new trucks by the president's children, using funds gifted to them, so that the trucks can be rented to the firm at the maximum reasonable lease rates.

6. Hiring the president's eldest son (16 years old) on a part-time basis with the most favorable compensation package allowed.

Figure B.2 (*continued*)

Appendix B

```
                        EASTWAY CONTRACTORS, INC.
                         PROJECT STATUS REPORT
                           JOBS IN PROGRESS
                              12/31/87
```

PROJECT	OFFICE CONDOS	TOWNHOUSES	APARTMENTS	RETAIL CENTER	OFFICE WAREHOUSE	TOTAL
Original Contract	2,000,000	1,500,000	1,200,000	2,200,000	1,500,000	8,400,000
Change Orders to Date	20,000	40,000	15,000	10,000	15,000	100,000
Adjusted Contract	2,020,000	1,540,000	1,215,000	2,210,000	1,515,000	8,500,000
Total Estimated Costs	1,625,000	1,300,000	1,110,000	1,840,000	1,125,000	7,000,000
Cost Incurred	1,000,000	750,000	700,000	350,000	450,000	3,250,000
Cost to Complete	625,000	550,000	410,000	1,490,000	675,000	3,750,000
Billed to Date	1,210,000	860,000	625,000	360,000	510,000	3,565,000
Unbilled Portion of Contract	810,000	680,000	590,000	1,850,000	1,005,000	4,935,000
Revenue Earned	1,250,000	900,000	775,000	420,000	600,000	3,945,000
Revenue to be Earned	770,000	640,000	440,000	1,790,000	915,000	4,555,000
(Overbilled),Underbilled	40,000	40,000	150,000	60,000	90,000	380,000
Total Estimated Gross Margin	395,000	240,000	105,000	370,000	390,000	1,500,000
Gross Margin Earned	265,000	140,000	65,000	70,000	155,000	695,000
Gross Margin to be Earned	130,000	100,000	40,000	300,000	235,000	805,000
Gross Margin %	19.55%	15.58%	8.64%	16.74%	25.74%	17.65%

```
= == == == == == == == == === == == == == == == == == == == == == == == == == == == == == == == == == == == =
```

```
                               JOBS TO START
                                 12/31/87
```

PROJECT:	OFFICE TOWNHOUSE	AUTO MALL	TENANT IMPROVEMENTS	RETAIL REMODEL	SELF STORAGE	TOTAL
Original Contract	580,000	1,200,000	400,000	600,000	700,000	3,480,000
Total Estimated Costs	470,000	1,008,000	332,000	498,000	581,000	2,889,000
Total Estimated Gross Margin	110,000	192,000	68,000	102,000	119,000	591,000
Gross Margin %	19.0%	16.0%	17.0%	17.0%	17.0%	17.0%

Figure B.2 (continued)

276

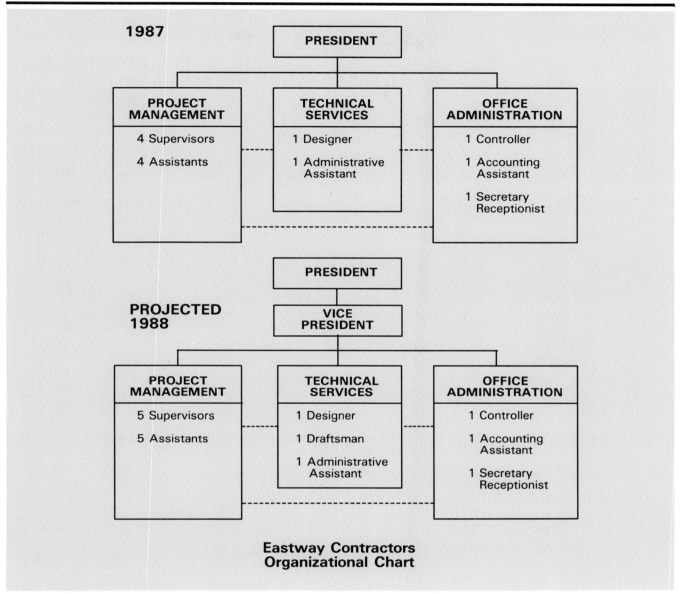

1987

PRESIDENT

PROJECT MANAGEMENT
- 4 Supervisors
- 4 Assistants

TECHNICAL SERVICES
- 1 Designer
- 1 Administrative Assistant

OFFICE ADMINISTRATION
- 1 Controller
- 1 Accounting Assistant
- 1 Secretary Receptionist

PRESIDENT

VICE PRESIDENT

PROJECTED 1988

PROJECT MANAGEMENT
- 5 Supervisors
- 5 Assistants

TECHNICAL SERVICES
- 1 Designer
- 1 Draftsman
- 1 Administrative Assistant

OFFICE ADMINISTRATION
- 1 Controller
- 1 Accounting Assistant
- 1 Secretary Receptionist

**Eastway Contractors
Organizational Chart**

Figure B.2 (continued)

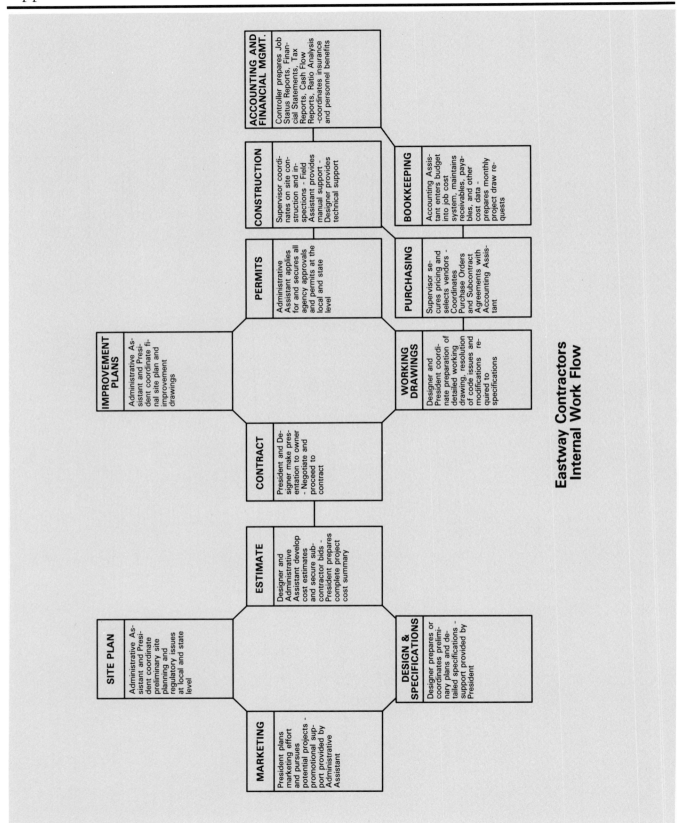

Figure B.2 (continued)

EASTWAY CONTRACTORS, INC.
1988 BUDGET SUMMARY
RELEVANT RANGE $7-$10 MILLION

| | | | | | DIRECT COST ALLOCATION | | | | | | | | | |
		% OF REVENUE		MARKUP	MATERIAL %	AMOUNT	LABOR %	AMOUNT	SUBCONTRACT %	AMOUNT	EQUIPMENT %	AMOUNT	MISCELLANEOUS %	AMOUNT
1	PROJECTED REVENUE		$8,000,000											
2	DIRECT COST	86.00%	6,880,000	100.00%	20%	1,376,000	3%	206,400	75%	5,160,000	1%	68,800	1%	68,800
3	GROSS PROFIT	14.00%	1,120,000	16.28%										
4	OPERATING COSTS													
5	INDIRECT COSTS						INDIRECT COST ALLOCATION							
6	Field Employee Benefits		47,500		20%	9,500	20%	9,500	40%	19,000	10%	4,750	10%	4,750
7	Field Equipment		24,929		20%	4,986	50%	12,464	10%	2,493	10%	2,493	10%	2,493
8	Field Insurance		34,000		20%	6,800	3%	1,020	75%	25,500	1%	340	1%	340
9	Field Vehicles		29,000		20%	5,800	30%	8,700	40%	11,600	5%	1,450	5%	1,450
10	Shop Maintenance		6,000		30%	1,800	40%	2,400	0%	0	20%	1,200	10%	600
11	Warranty		24,000		20%	4,800	3%	720	75%	18,000	1%	240	1%	240
12	Misc. Indirect Expenses		16,000		20%	3,200	3%	480	75%	12,000	1%	160	1%	160
13	TOTAL INDIRECT COSTS	2.27%	181,429	2.64%	2.68%	36,886	17.10%	35,284	1.72%	88,593	15.45%	10,633	14.58%	10,033
14	GENERAL & ADMINISTRATIVE COSTS													
15	Business Insurance		10,000											
16	Occupancy		56,857											
17	Office Employees		200,357											
18	Office Equipment		11,700											
19	Office Supplies		15,000											
20	Office Vehicles		19,757											
21	Officer Compensation		189,500											
22	Professional Fees		15,714											
23	Sales & Marketing		47,000											
24	Travel & Entertainment		14,286											
25	Misc. Gen. & Admin. Expenses		45,429											
26	TOTAL GEN. & ADMIN. COSTS	7.82%	625,600	9.09%										
27	TOTAL OPERATING COSTS	10.09%	807,029	11.73%										
28	OPERATING PROFIT	3.91%	312,971	4.55%										
29	ALL OTHER INCOME & EXPENSE (NET)	0.10%	8,000	0.12%										
30	NET PROFIT BEFORE TAX	4.01%	$320,971	4.67%										

Figure B.2 (continued)

```
                    EASTWAY CONTRACTORS, INC.
                    PRO FORMA BALANCE SHEET
                    DECEMBER 31, 1988 WITH
                COMPARATIVE ACTUAL DECEMBER 31, 1987

                             ASSETS
                             ------
```

	PROJECTED 1988	ACTUAL 1987
Current Assets		
Cash	$325,000	$295,000
Accounts Receivable – Trade	510,000	400,000
Accounts Receivable – Retention	105,000	80,000
Inventory	10,000	10,000
Costs and estimated earnings in excess of billings	75,000	50,000
All Other Current Assets	25,000	25,000
Total Current Assets	1,050,000	860,000
Plant, Property and Equipment – at cost		
Construction Equipment	65,000	40,000
Trucks and Autos	46,000	30,000
Office Equipment	46,000	30,000
	157,000	100,000
Less: Accumulated depreciation and amortization	45,000	15,000
Plant, Property and Equipment, net	112,000	85,000
Other Assets		
Joint Ventures	0	0
Other Non-Current Assets	40,000	40,000
Total Other Assets	40,000	40,000
TOTAL ASSETS	$1,202,000	$985,000

Figure B.2 (continued)

```
              LIABILITIES AND STOCKHOLDER'S EQUITY
              ----------------------------------------
                                        PROJECTED       ACTUAL
                                          1988            1987
                                          ----            ----

Current Liabilities
  Notes Payable                         $40,000         $30,000
  Accounts Payable - Trade              201,000         265,000
  Accounts Payable - Retention           40,000          25,000
  Billings in excess of costs and
    estimated earnings                   86,000          60,000
  Income Taxes Payable                   45,000          25,000
  Current Maturity - Long Term Debt      15,000          15,000
  All Other Current Liabilities          13,029          25,000
                                      ----------      ----------

    Total Current Liabilities           440,029         445,000

Long Term Debt                           40,000          30,000
Deferred Income Tax                      20,000          25,000
Other Non-Current Liabilities            15,000          15,000
                                      ----------      ----------

                                         75,000          70,000

    Total Liabilities                   515,029         515,000


Stockholders' Equity
  Common Stock                           10,000          10,000


  Retained Earnings
    Balance - beginning of period       460,000         280,000
    Net Profit/(Loss)                   216,971         180,000
    Balance - end of period             676,971         460,000
                                      ----------      ----------

    Total Stockholders' Equity          686,971         470,000


    TOTAL LIABILITIES AND
      STOCKHOLDERS' EQUITY           $1,202,000        $985,000
                                      ==========      ==========
```

Figure B.2 (continued)

EASTWAY CONTRACTORS, INC.
PROFORMA INCOME STATEMENT
FOR THE YEAR ENDING DECEMBER 31, 1988
COMPARATIVE ACTUAL FOR THE YEAR ENDED DECEMBER 31, 1987

	PROJECTED 1988		ACTUAL 1987	
	AMOUNT	% OF INCOME	AMOUNT	% OF INCOME
REVENUE				
Construction Contracts	$7,928,000	99.10%	$5,950,000	99.17%
Design Contracts	72,000	0.90%	50,000	0.83%
TOTAL REVENUE	8,000,000	100.00%	6,000,000	100.00%
COST OF REVENUE				
DIRECT COST	6,880,000	86.00%	5,140,000	85.67%
INDIRECT COSTS				
Field Employees	47,500	0.59%	20,000	0.33%
Field Equipment	24,929	0.31%	16,000	0.27%
Field Insurance	34,000	0.43%	25,000	0.42%
Field Vehicles	29,000	0.36%	25,000	0.42%
Warranty	24,000	0.30%	20,000	0.33%
Other Indirect Expenses	16,000	0.20%	24,000	0.40%
TOTAL INDIRECT COSTS	175,429	2.19%	130,000	2.17%
TOTAL COST OF REVENUE	7,055,429	88.19%	5,270,000	87.83%
GROSS PROFIT	944,571	11.81%	730,000	12.17%
OPERATING EXPENSES				
Business Insurance	10,000	0.13%	7,000	0.12%
Occupancy	56,857	0.71%	45,000	0.75%
Office Employees	200,357	2.50%	140,000	2.33%
Office Equipment	11,700	0.15%	16,000	0.27%
Office Supplies	15,000	0.19%	15,000	0.25%
Office Vehicles	19,757	0.25%	15,000	0.25%
Officer Compensation	189,500	2.37%	150,000	2.50%
Professional Fees	15,715	0.20%	12,000	0.20%
Sales & Marketing	47,000	0.59%	35,000	0.58%
Travel & Entertainment	14,285	0.18%	10,000	0.17%
Other Gen. & Admin. Expenses	45,429	0.57%	35,000	0.58%
TOTAL OPERATING EXPENSES	625,600	7.82%	480,000	8.00%
OPERATING PROFIT	318,971	3.99%	250,000	4.17%
ALL OTHER INCOME/(EXPENSE)-NET	8,000	0.10%	10,000	0.17%
PROFIT BEFORE TAX	326,971	4.09%	260,000	4.33%
INCOME TAX	104,000	1.30%	80,000	1.33%
NET PROFIT	$222,971	2.79%	$180,000	3.00%

Figure B.2 (continued)

EASTWAY CONTRACTORS, INC.
ANNUAL CASH FLOW BUDGET
JANUARY 1, 1988 - DECEMBER 31, 1988

	TOTAL	JANUARY	FEBRUARY	MARCH	APRIL	MAY	JUNE	JULY	AUGUST	SEPTEMBER	OCTOBER	NOVEMBER	DECEMBER
BEGINNING BALANCE		85,000	165,900	27,900	29,950	35,600	38,200	31,000	25,800	32,800	77,850	21,800	31,900
NET JOB CASH FLOWS													
In Process:													
Job #1	205,100	79,700	(70,300)	36,300	(88,900)	600	100	8,300	71,300	168,000	0	0	0
Job #2	120,000	120,000	0	0	0	0	0	0	0	0	0	0	0
Job #3	149,500	(48,000)	55,000	48,000	75,000	(22,000)	1,500	40,000	0	0	0	0	0
Job #4	(11,100)	0	(50,000)	25,000	48,000	2,000	500	2,500	(40,000)	(35,000)	20,000	15,000	900
Job #5	(25,300)	(14,000)	(100,000)	15,000	(2,000)	1,500	65,000	(800)	10,000	0	0	0	0
Projected New Jobs:													
Job #6	65,800	0	0	0	0	0	20,000	40,000	(3,000)	(700)	(5,800)	300	15,000
Job #7	82,500	0	(500)	0	55,000	(21,000)	14,000	15,000	400	1,600	10,000	8,000	0
Job #8	30,000	0	0	0	0	0	0	0	0	0	(25,000)	(10,000)	65,000
TOTAL NET JOB CASH FLOWS	616,500	137,700	(165,800)	124,300	87,100	(38,900)	101,100	105,000	38,700	133,900	(800)	13,300	80,900
TOTAL CASH AVAILABLE		222,700	100	152,200	117,050	(3,300)	139,300	136,000	64,500	166,700	77,050	35,100	112,800
EXPENDITURES													
OPERATING EXPENSES													
Indirect Costs	279,000	21,000	23,000	26,000	22,000	23,000	24,000	25,000	24,000	27,000	21,000	20,000	23,000
Gen. & Admin. Costs (less depr.)	414,000	36,000	34,000	36,000	34,000	35,000	34,000	35,000	33,000	35,000	34,000	33,000	35,000
OTHER INCOME & EXPENSES (Net)	(3,100)	(600)	(200)	(150)	50	100	(100)	(200)	(700)	(550)	(250)	(300)	(200)
CAPITAL EXPENDITURES	10,000	0	0	0	0	0	0	0	0	10,000	0	0	0
CURRENT PORTION LONG TERM DEBT	5,100	400	400	400	400	400	400	400	400	400	500	500	500
TOTAL CASH DISBURSEMENTS	705,000	56,800	57,200	62,250	56,450	58,500	58,300	60,200	56,700	71,850	55,250	53,200	58,300
ENDING CASH BALANCE		165,900	(57,100)	89,950	60,600	(61,800)	81,000	75,800	7,800	94,850	21,800	(18,100)	54,500
DEBT ADJUSTMENTS													
LONG TERM DEBT INCREASE (DECREASE)	8,000	0	0	0	0	0	0	0	0	8,000	0	0	0
SHORT TERM DEBT INCREASE (DECREASE)	20,000	0	85,000	(60,000)	(25,000)	100,000	(50,000)	(50,000)	25,000	(25,000)	0	50,000	(30,000)
TOTAL DEBT ADJUSTMENT INC.(DEC.)	28,000	0	85,000	(60,000)	(25,000)	100,000	(50,000)	(50,000)	25,000	(17,000)	0	50,000	(30,000)
ADJUSTED ENDING CASH BALANCE		165,900	27,900	29,950	35,600	38,200	31,000	25,800	32,800	77,850	21,800	31,900	24,500
CUMULATIVE DEBT													
LONG TERM DEBT													
Beginning Balance	40,000	40,000	39,600	39,200	38,800	38,400	38,000	37,600	37,200	36,800	44,400	43,900	43,400
Increase(Decrease)	2,900	(400)	(400)	(400)	(400)	(400)	(400)	(400)	(400)	7,600	(500)	(500)	(500)
ENDING LONG TERM DEBT	42,900	39,600	39,200	38,800	38,400	38,000	37,600	37,200	36,800	44,400	43,900	43,400	42,900
SHORT TERM DEBT													
Beginning balance	0	0	0	85,000	25,000	0	100,000	50,000	0	25,000	0	0	50,000
Increase(Decrease)	20,000	0	85,000	(60,000)	(25,000)	100,000	(50,000)	(50,000)	25,000	(25,000)	0	50,000	(30,000)
ENDING SHORT TERM DEBT	20,000	0	85,000	25,000	0	100,000	50,000	0	25,000	0	0	50,000	20,000

Figure B.2 (continued)

<u>EASTWAY CONTRACTORS</u>

<u>FIXED COST CONTRACT</u>

This AGREEMENT, made this _____ day of _____, 19 ___, by and between Eastway Contractors (hereinafter called "Builder") and _____ (hereinafter called "Owner") shall include all the services described herein for the construction or re-modeling of _____ (hereinafter called "Project"). The documents which supplement this Agreement shall consist of the plans prepared by _____ and dated _____, the attached Specifications, the attached Construction Standards and Warranty, and the attached AIA Documents G702 and G703, which outline the schedule of values and billing format.

I. CONTRACT SUM/PAYMENTS

    A. In consideration of the Agreement and payments herein set forth, subject to ratification by both parties no later than _____, the Builder agrees to furnish and the Owner agrees to purchase all services, materials, and labor required to complete the project in accordance with the attached documents and the quality standards typical in the local industry. Following ratification, either party may declare this contract null and void if all applicable approvals, including financing at terms satisfactory to the Owner, are not secured such that construction may begin by _____. The total purchase price to be paid by Owner to Builder for all items, subject to additions and deductions by Change Order, is _____ _____ Dollars ($_____), which shall be paid according to the schedule of payments and retentions shown on the attached AIA form G703.

    B. The Owner agrees to provide written documentation from his lender as the availability of funds required to satisfy all payments covered in this contract. The builder shall submit draw requests approximately every _____, and the Owner agrees to pay such draw requests within _____ days after receipt, with interest accruing at the rate of _____% on any sums unpaid without cause beyond the due date. The draw request for work completed as well as materials suitably stored on site shall be prepared upon IAI forms G702 and G703. Any retentions shall be placed

Figure C.1

into a mutually agreed upon interest bearing escrow account, will all interest accruing to the Builder. Retention monies shall be released to the Builder within _____ of substantial completion. If any payment becomes overdue by 7 days, following written notice, the Builder may, without prejudice to any other remedy he may have, stop work and recover from the Owner payment for all work executed, as well as the associated operating expenses and lost profits on all completed work. In the event of a dispute pertaining to a portion of any draw, only sums reasonably representative of the portion in dispute shall be withheld.

C.   In the event that the Builder is unable for any reason beyond his control, including but not limited to unforeseen ground conditions, to build and convey the structure provided for in this contract at the price stipulated, any payments will be refunded, less any advances and operating expenses incurred on the account of the Owner prior to the discovery of any conditions which cause the Builder to be unable to perform the contract. If the Builder fails or neglects to carry out the work in accordance with the Contract Documents, or otherwise to perform in accordance with this Agreement, and fails within seven days after receipt of written notice to commence and continue correction of such default, the Owner may terminate this Agreement and finish the work by whatever method he may deem expedient. If the unpaid balance of the contract sum exceeds the expense of finishing the work, such excess shall be paid to the Contractor. If such expense exceeds such unpaid balance, the Contractor shall pay the difference to the Owner.

II. Completion

A.   The Builder agrees to substantially complete (suitable for occupancy and issuance of Use and Occupancy Certificate, but prior to completion of all punch list items) the Project within _____ from the start of construction, subject, however, to any causes beyond the control of the Builder which may prevent or delay such completion, including without limitation any of the following: weather, ground conditions, Change Orders, arbitration, availability of materials, governmental rulings, acts of God, issuance of permits, or litigation pertaining to any of the foregoing. The Owner agrees to compensate the Builder for additional costs incurred resulting from postponements in construction imposed by and for the benefit of the Owner, e.g. delays due to slow sales or rentals in a speculative project or delays due to the unavailability of specialty products selected by the owner subsequent to this agreement.

B.   Prior to final payment, the Builder shall provide, if requested, an Affidavit to the Owner stating that the Builder has paid all valid bills and charges for

Figure C.1 (continued)

materials, labor, or otherwise in connection with or arising out of the construction of the project, and will hold the Owner free and harmless against any claims of lien for labor or materials filed against the property. Upon substantial completion, Builder and Owner shall make a final inspection, listing all items found incomplete or defective on an inspection form. All final payments shall become due and payable to Builder when all items on the inspection form are completed or corrected.

C. The Owner agrees not to move any materials or equipment into the structure until it has been completed and the total purchase price has been paid in full, unless agreed to otherwise in writing. Failure of any inspecting agency or financial institution to issue a final approval for any reason beyond the control of the Builder shall not in any way prejudice the Builder's rights to monies due or to file a mechanics lien for work performed or materials furnished.

III. ALLOWANCES/CHANGE ORDERS

A. The term "allowance," when used in this contract, shall be interpreted as the amount allowed by the Builder to supply and install the materials referred to, unless otherwise specified, when the costs may be of an indefinite amount because of optional selections by the Owner and/or other causes beyond the control of the Builder. If a selection is made which results in a total direct cost which is less than the allowance figure, the savings shall be deducted from the draw in which that allowance item is covered. If a selection is made which results in a total direct cost which exceeds the allowance figure, the Owner agrees to pay in advance the additional direct costs as well as _____% to cover administrative expenses, taxes, insurance, accounting, warranties, other operating expenses, and profit. All selections shall be completed and submitted to Builder in writing within forty-five (45) days of contract ratification.

B. The Owner may issue additional instructions, request additional work, or direct omissions of work included in the contract documents. No change, except for costs incurred in excess of specified allowances, shall be legally binding upon either party until an approved Change Order form is returned to the Builder with proper signatures and payment. Any additional direct costs for any Change Order shall be increased by the percentage for operating expenses and profit described in Article IIIA. All Change Orders, including those representing a net credit or no additional direct cost, may carry an administrative charge of $_____ at the option of the Builder. Any credit for omissions may be reduced to the extent the Builder incurs

Figure C.1 (continued)

order cancellation charges or labor and overhead expended in the performance of the contract, which is to no avail due to the Change Order.

C. In the event any unforeseen condition is encountered in the performance of this Project, including but not limited by such things as rock, other unforeseen ground conditions, or defects in existing structures, which requires additional labor or materials, the additional direct cost associated with such work along with markup for operating expenses and profit as described in Article IIIA, shall be described and submitted to the Owner on a written Change Order and shall be fully paid upon execution of such work.

D. Any labor or materials which are to be the responsibility of the Owner as set forth in the contract documents, and therefore, not within the control of the Builder, shall be planned and controlled by the Owner according to the Builder's normal scheduling procedures and standards of quality. This shall include any allowance item for which the owner opts to provide labor or to purchase materials. Any such materials supplied by the Owner shall not be installed by any Builder employee or Subcontractor unless otherwise agreed and shall carry no warranty of any kind. All materials and equipment remaining on site and not required for the completion of the Project, other than those covered by allowance or provided by the Owner, are the sole property of the Builder and shall be removed unless otherwise agreed.

IV. GENERAL

A. Unless otherwise specified, the Owner agrees that during the entire time construction is in process, the Builder, his suppliers, and his subcontractors shall have access to and across any portion of the site for any use required by the Builder's operation, including the placement of advertising signs. The Builder agrees to exercise due care in all on-site operations, however, unless specified, shall not be responsible for unavoidable damage to existing site appurtenances during the normal performance of work by Builder employees, subcontractors, or suppliers. Any damage for which Builder shall be responsible shall be described in the specifications as to location, condition, and degree of responsibility.

B. The Builder agrees to comply with all local, state, and federal rules and regulations bearing on the performance of this agreement, and to take all safety precautions required by the safety agencies governing the local jurisdiction. The Builder warrants that it is covered by Comprehensive General Liability Insurance in the amount of $_____ and the statutory levels of Workers Compensation. The Builder shall provide Builders

Figure C.1 (continued)

Risk Insurance in the amount of $_____ until substantial completion, at which time the owner shall be fully responsible for insuring the project. The Builder agrees to indemnify and hold the Owner harmless in the event of any claims, damages, losses, or expenses resulting from any negligent act of the Builder, his subcontractors, or his associates.

C. Until the contract is complete and all monies paid, the Owner shall not enter into any written or verbal contractual agreement with any Builder employee or Subcontractor for services or materials of any type. All communications regarding any contract document, Change Order, or issues of a technical nature shall be directed only to personnel identified by the Builder as a qualified representative.

D. The parties hereto mutually covenant and agree that should a dispute arise regarding this contract or any work performed in connection with it, said dispute shall be submitted to arbitration in accordance with the Construction Industry Arbitration Rules of the American Arbitration Association. The party whose position is substantially favored by the arbitrators shall have its legal, direct, indirect, administrative, and all other expenses paid for by the other party.

E. This contract, and the other documents described herein, contain the final and entire agreement between the parties hereto and neither they nor their agents shall be bound by any terms, conditions, or representations discussed in previous negotiations but not written herein; and it shall inure to the benefit and be binding upon the heirs, personal representatives, successors, and assigns of the parties hereto.

CONTRACTOR   OWNER

BY:_____   BY:_____

DATE:_____   DATE:_____

Figure C.1 (continued)

GLOSSARY

Accelerated cost recovery system (ACRS)
A method of deducting tax to recover the cost of depreciable personal property.

Accelerated methods
Depreciation methods that result in larger depreciation expense during the early years of asset life, with gradually decreasing depreciation expense in later years.

Account
A recording device used for sorting accounting information into similar groupings.

Accounting
The set of rules and methods by which financial and economic data are collected and transformed into useful reports for decision making.

Accounting cycle
The steps that must be followed to process and record information, summarize and classify this information, and prepare the records to accomplish these steps during the next period.

Accounting system
The various processing steps, equipment, personnel, and procedures that change original data to useful form.

Accounts payable
Unpaid amounts or debts a business owes to creditors from purchases on open account.

Accounts receivable
Amounts due from customers for sales or services rendered.

Accrual basis of accounting
The method that assumes that revenue is realized at the time of the sale of goods or services, regardless of when the cash is received; expenses are recognized at the time the services are received and utilized or an asset is consumed in the production of revenue, regardless of when payment for these services or assets is made.

Accruals
A classification of adjustments that includes accrued revenues and accrued expenses.

Accrued
Accumulated or grown over a period of time.

Accrued expenses
Expenses that have been incurred in a given period—for example, services received and used—but that have not yet been paid or recorded.

Accrued liability
The liability for an expense that has been accumulated but not paid or recorded.

Accrued revenues
Revenues that have been earned in a given period but that have not yet been collected or recorded.

Accumulated Depreciation
An account that reveals all past depreciation that has been recorded on a depreciable property, plant, and equipment item and charged against revenue; it is, in essence, a postponed credit to the applicable property, plant, and equipment account.

Adjusted cash balance
The true cash balance resulting from reconciling the difference between the balance reported by the bank and the amount shown on the depositor's books.

Adjusting entries
Entries for regular, continuous transactions the recording of which has been postponed to the end of an accounting period for the convenience of the account; they are made to update revenue, expense, asset, liability, and owner's equity accounts as required by the accrual basis of accounting.

Aging
A method of classifying individual receivables by age groups, according to the time elapsed from the due date.

Aging schedule
A columnar work sheet showing the individual receivables by age groups, according to time elapsed from the due date. The individual age groups are also totalled, and a percentage analysis is computed to aid in determining the allowance for doubtful accounts.

Amortization
Often used as a general term to cover write-down of assets; it is most commonly used to describe periodic allocation of the costs of intangible assets, to expense.

Annual effective interest rate
The correct interest rate computed on only the remaining balance of an unpaid debt for the specific time period, usually stated as an annual fraction.

Asset
An item of value owned by an economic enterprise.

Auditing
An independent review of an entity's accounting reports, usually made by a certified public accountant.

Bad Debts Expense
An expense account showing the estimated uncollectible credit sales made in a given time period (for one year if the accounting period is a year), or actual write-offs if the direct write-off method is used.

Bad Debts Recovered
A revenue account that is credited for the recovery of an account receivable previously written off under the direct write-off method.

Balance
The difference between the total of the debit amounts and the total of the credit amounts in an account.

Balance sheet
The statement that summarizes the assets, liabilities, and owner's (or owners') equity of a business unit as of a specific date.

Book value
The net amount at which an asset is carried on the books or reported in the financial statements; it is the asset's cost at acquisition, reduced by the amount of accumulated depreciation on the asset. Also called carrying value and undepreciated cost.

Break-even point
The volume of sales at which the business will neither earn income nor incur a loss.

Budget
A financial plan for a future period developed in organizational detail.

Capital budgeting
The allocation and commitment of funds to long term capital investment projects.

Capital expenditures
Payments or promises to make future payments for assets that will benefit more than one accounting period. They are carried forward as assets.

Capitalize
To increase the property, plant, and equipment book value with expenditures that increase the EUL or the valuation of a property, plant, and equipment asset.

Cash
Currency, coins, travelers' checks, checks, and any other items that a bank will accept for deposit.

Cash basis of accounting
The recognition of revenue at the time that cash is received for the sale of goods and services and the recognition of expenses when paid.

Cash forecast
A subdivision of the budget structure, this schedule sets forth a prediction of monthly cash inflows and outflows.

Cash payments journal
A special journal in which all cash payments are recorded.

Cash receipts journal
A special journal in which all cash receipts are recorded.

Chart of accounts
A list of all accounts in the general ledger. Their numbering system indicates the types of accounts by sub-groups.

Closing entries
Those journal entries that close the nominal accounts at the end of a period—that is, reduce these accounts to a zero balance. In the closing process, the net effect of this closing is transferred to the owner's capital account.

Closing the books
The process of clearing the temporary or nominal accounts at the end of a period; this requires preparation of closing journal entries and posting of these entries to the nominal accounts that are closed.

Comparative balance sheets
The balance sheet as of a given date compared with one or more immediately preceding balance sheets.

Compound entry
A journal entry with more than one debit or credit.

Controlling
The management function of monitoring actual versus planned activity and taking corrective action where appropriate.

Controlling account
One account in the general ledger that controls and is supported by a group of accounts in separate subsidiary ledgers.

Corporation
A form of business which is a legal entity as well as an accounting entity.

Cost
The amount paid or payable in either cash or its equivalent for goods, services, or other assets purchased.

Cost allocation
The process of apportioning costs among functions.

CPA
A certified public accountant.

Credit

The right-hand side of the "T" form of an account, the amount shown on the right side of an account, or the process of placing an amount on the right side of an account.

Creditors

Persons or groups to whom debts are owed.

Current assets

Cash and other assets that will be consumed or converted into cash within one year.

Current liabilities

Liabilities to be paid within one year.

Debit

The left-hand side of the "T" form of an account, the amount shown on the left side of an account, or the process of placing an amount on the left side of an account.

Depreciation

The allocation of a part of the cost of a property, plant, or equipment item (that has a limited useful life) over its estimated useful life.

Direct Costs

The labor, material, subcontractor, and heavy equipment costs directly incorporated into the construction of physical improvements.

Disbursement

An actual payment by cash or check.

Double-declining balance method (DDB)

An accelerated depreciation method in which a constant rate—twice that of the straight line rate—is applied to compute annual charges.

Double-entry accounting

A system of recording both the debit and credit aspect of each transaction.

Employee's Withholding Allowance Certificate

The form (W-4) prepared by each employee showing the number of withholding allowances currently claimed for establishing the amount of federal income taxes to be withheld by the employer from the employee's earnings.

Employer's Quarterly Federal Tax Return

The form (941) on which employers must report the amount of federal income taxes and FICA taxes withheld and the dates and amounts deposited during each quarterly period.

Entity

The focal point of attention of accounting records; an organization such as a business.

Estimated useful life (EUL)
An estimate, made at the time of acquisition, of the term of usefulness of an asset (may be in years, working hours, or units of output).

Expense
Expired cost; the material used and service utilized in the production of revenue during a specific period.

Fair market value
Value determined by informed buyers and sellers based usually on current invoice or quoted prices.

Federal income tax withholding
An amount deducted from gross pay by the employer and remitted to the Internal Revenue Service.

Federal Insurance Contributions Act (FICA)
A federal law requiring both employers and employees to contribute equal amounts based on a stated percentage of taxable wages paid; it provides funds to the Social Security program.

Federal Unemployment Tax Act (FUTA)
A federal law that levies a tax on the employer at a specified rate up to a limited amount of wages paid; it provides funds to the Social Security program.

Fixed costs
These costs can be defined in two different ways. In total, fixed costs remain the same as volume changes. On a per unit basis, fixed costs decline as volume increases. Examples include office rent, administrative salaries, insurance, and office supplies.

Flexible budget
A budget that gives recognition to varying levels of volume and the costs that change with these levels. A series of fixed budgets for various levels of production.

Funds
In the context of the statement of changes in financial position, funds may mean cash, working capital, cash and temporary investments in marketable securities, current assets, or all financial resources, depending upon the needs of the user.

General and administrative expenses
The cost of goods or services generally reflecting the cost of operations.

General journal
The book of original entry in which all transactions that do not fit into special journals are recorded.

General ledger
The main group of ledger accounts incorporated into the trial balance. It does not include the separate accounts that are in the subsidiary ledgers.

Generally accepted accounting principles (GAAP)
A body of guidelines that are followed because of broad acceptance by the accounting profession.

Gross margin
The excess of net sales revenue over direct costs.

Gross pay
Total wages before any deductions; this amount is the salaries and wages expense.

Income statement
A statement showing all revenue and expense items for a given period, arranged so that total expenses are subtracted from total revenues, revealing the net income earned during that period.

Indirect Costs
Costs incurred through field production, but not directly incorporated into the construction of physical improvements. Indirect costs are typically difficult to track and attribute to specific projects and are, therefore, allocated to all projects. Examples include: field personnel benefits, field insurance, field vehicles used on multiple projects, and warranty.

Inflation
An upward change in the general price level resulting in a decrease in the purchasing power of the monetary unit.

Intangible assets
Nonphysical assets the ownership of which is expected to yield benefits.

Interest
The price of credit; a rental charge for the use of money.

Interim statements
Any statements that are made during the period, but not including those statements made at the end of the period.

Internal control
The plan of organization, procedures, and equipment used in a business to protect its assets from improper use and to promote efficiency in operations.

Invoice
A form that provides evidence of the sale and delivery of merchandise; it indicates that an account receivable exists.

Job cost accounting system
An accounting system in which period-to-period cost information is developed for each project.

Journalizing
The process of recording a transaction, in terms of its debits and credits, in a record referred to as a *journal*.

Lease
The right to use, over a fixed period of time, property belonging to others.

Ledger
The book that contains all the ledger accounts; or a collection of ledger accounts in any form.

Liability
An obligation of a business, or a creditor's claim against the assets of a business.

Liquidity
The nearness to cash of assets and liabilities.

Long-term liabilities
Debts of a business that are not due for at least one year.

Loss
Expired cost; materials and services utilized that did not produce any revenue and that result from the incidental transactions of an entity.

Matching
Expenses incurred in the generation of revenues should be matched against those revenues to determine *income*.

Matching concept
The matching of incurred expenses and earned revenue for a given time period to determine net income for that period.

Net assets
Total assets minus total liabilities.

Net income
Excess of revenue over expenses for a given period.

Net loss
Excess of expenses over revenue for a given period.

Net pay
Wages after all deductions; this is referred to as *take-home pay*.

Noncurrent account
Any account on the balance sheet other than current assets and current liabilities—specifically, any one of the long term investments; property, plant, and equipment; intangibles; long term liabilities; or stockholders' equity accounts.

Notes payable
A balance sheet caption most commonly used for short term notes to creditors.

Notes receivable
Claims against individuals or companies supported by formal written promises to pay; a note receivable may be either a trade note or a nontrade note.

Operating expenses
The combination of indirect and general and administrative expenses.

Other revenue and other expenses
Items of ordinary revenue and expense that arise from a source other than the basic business of the company.

Owner's equity
The owner's or owners' claims against the assets of a business. As used in this text, owner's equity implies that the business is a single proprietorship and, therefore, represents the proprietor's claims against assets of the single proprietorship.

Payroll deductions
Amounts withheld from gross pay by the employer; these include federal and state taxes, union dues, and medical insurance.

Percentage of completion method
The revenue realization method, appropriate for large construction projects, which recognizes revenue in proportion to the percentage of estimated total costs that have been incurred.

Planning
Setting the goals and objectives for a future period.

Posting
The process of transferring an amount recorded in the journal to the indicated account in the ledger.

Prepaid items
Unconsumed amounts of current assets that will normally be used in the coming year.

Present value method
A means of converting projections of cash inflows and outflows over time to their present value, using an estimated discounting rate, to evaluate capital expenditures.

Processing
A method of sorting and analyzing data in terms of their effect on accounts; processing is done through a journal and various ledgers.

Profit
The reward to an organization for rendering services or providing products.

Property, plant, and equipment
Assets the use of which will provide benefits over more than one accounting period; these include tangible assets (land, buildings, machinery) and wasting assets (oil, gas, minerals).

Purchase order
A formal written authorization to a vendor to provide certain goods or services and to bill the buyer for them at the specified price. The purchase order becomes a contract when it is accepted by the vendor.

Purchases journal
A special journal in which credit purchases of merchandise are recorded.

Ratio analysis
The analysis of the proportion of financial figures from period to period expressed in a percentage, a decimal, or a fraction.

Relevant volume range
The volume range over which cost behavior is somewhat stable and can be reasonably predicted.

Replacement cost
The current cost of replacing one specific asset with another asset of equivalent capacity.

Revenue
A term describing the inflows of assets received in exchange for services rendered, sales of products or merchandise, earnings from interest and dividends on investments, and advantageous settlement of liabilities at less than the amount of the debt.

Reversing entries
Entries made on the first day of a new fiscal period that are reversals of adjusting entries.

Sales
An account credited for the value of goods or services.

Sales journal
A special journal in which credit sales of merchandise are recorded.

Schedule of accounts payable
A listing of a business' individual creditors with the amount owed to each and the total owed to all creditors at a given moment in time.

Schedule of accounts receivable
A listing of a business' individual customers (debtors) with the amount owed by each and the total amount receivable from all customers.

Semi-variable costs
Costs that possess both a fixed component and a variable component.

Single proprietorship
A business owned by a single individual.

Source documents
Business papers on which each individual transaction is recorded. They provide objective evidence of transactions.

Special journals
Books of original entry that receive the initial recording of specialized classifications of transactions.

Statement classification
Grouping of similar elements of the accounting equation in a financial statement.

Statement of owner's equity
A statement showing the changes that occurred in a proprietor's capital during a given period.

Subsidiary ledger
A group of accounts in a separate ledger that provides information in detail about one controlling account in the general ledger.

"T" account
A simple form of ledger account in the shape of a "T", used for analyzing transactions and for teaching purposes.

Trade discount
A percentage reduction in a list price that results in the net price or billing price. Unlike the cash discounts, it is not recorded in the accounts.

Trade receivable
A claim against a customer arising from the sale of either merchandise or ordinary services.

Transaction
A business activity or event.

Trial balance
A statement that shows the name and balance of all ledger accounts arranged according to whether they are debits or credits. The total of the debits must equal the total of the credits in this statement.

Variable costs
Variable costs change directly in proportion to changes in volume.

Variance
The difference—favorable or unfavorable—between actual and planned performance.

Voucher
A serially numbered written authorization for each expenditure. It is prepared from the documents that serve as evidence of the liability.

Voucher register
A columnar journal for recording and summarizing all liabilities approved for payment.

Voucher system
A method of accumulating, verifying, recording, and disbursing all the expenditures of a business. It covers all payments except those from the petty cash fund.

Working capital
Current assets less current liabilities, or the amount of current assets not required to liquidate current liabilities.

Work-in-process
The inventory of partly finished projects in various stages of completion at any given time.

INDEX

power to sever or aggregate contracts, 209-210
Inventory
  computerized, 240
  excess, 138
  types, 147-148
Inventory management, 146-147
Investment decisions
  capital expenses, 166
Investment plans, 97
  risks, types, 140
  services, 141
Investments
  short-term, 139-141
  strategy trends, 11
Invoices, 29

J

Job cost data, 28
Job cost ledger, 146, 148
  computerized, 234-235
Job cost processing, 34
Job cost system, 35, 48-56
Job status report, 51-56
Journals, 29-31

L

Labor costs
  direct, 148
  indirect, 148
Leases
  evaluating, 188-189
  types, 187-188
Leasing, 187-189
Ledgers, 31
Legal actions, avoiding, 9
Lenders. *See* Banks; Credit, trade
Leverage ratios, 74-75
Liabilities, 28
  current, 39, 42
  long-term, 39, 42-43
Liquidation, 197
Liquidity, 67-74
  current ratio, 67-72
  risk, 140, 141
Litigation
  contractor's credit image, 183
Loan application, 185-187
Loans
  discounted, 178
  See also Financing
  short-term, 176-177
  term, 179-181
Loan terms, 184
Lock box, 135
"Look Back" Rule, 205
Losses
  passive, 197-198, 215

M

Management Advisory Services (MAS), 19
Management Information System Mission Statement, 227

Management Information Systems, 223-224
Managerial trends, 11
Marketing
  factor in business trends, 83
  inadequate, 14
  trends, 10
Markup, 99
Materials
  excess, 138
  timing of purchases, 137
  See also Inventory
Maturity matching, 179
MIS. *See* Management Information Systems
Modem, 224
Money market funds, 139
Multiple discriminant analysis, 64

N

Net present value, 163
Networking, 225
Net worth to fixed asset ratio, 75

O

Officers' compensation
  contractor's credit image, 183
  effect on financial analysis, 65
  unreasonable, 198
Operating expenses
  contractor's credit image, 182
Operating lease, 188
Organization of business
  tax consequences, 194-200
  types, 194
Outside parties
  need for financial statements, 27, 31
Overbilling, 152
Overhead. *See* Costs, general and administrative
Owners
  relationship to architects and contractors, 10
  See also Officers

P

Partnership organization
  tax consequences, 195
Payables
  factor in business trends, 83
  See also Accounts payable
Payable turns, 73
Payroll
  computerized, 235-237
Payroll information, 31
Percentage of completion accounting, 36, 37-38
  modified, 203-204
Percentage of completion-capitalized cost accounting, 200, 205
Percentage of completion schedule. *See* Job status report
Planning and control process, 93-94
Planning, strategic, 94
Pledging, 176